IRRIGATION, AGRICULTURE AND THE RAJ: PUNJAB, 1887-1947

SOUTH ASIA INSTITUTE, NEW DELHI BRANCH
HEIDELBERG UNIVERSITY
SOUTH ASIAN STUDIES NO. XXX

IRRIGATION, AGRICULTURE AND THE RAJ: PUNJAB, 1887-1947

M. MUFAKHARUL ISLAM

MANOHAR
1997

First published 1997

ISBN 81-7304-167-9

Published by
Ajay Kumar Jain for
Manohar Publishers & Distributors
2/6 Ansari Road, Daryaganj
New Delhi 110 002

Typeset by
A J Software Publishing Co. Pvt. Ltd.
305 Durga Chambers, 1333 D.B. Gupta Road
Karol Bagh, New Delhi 110 005

Printed at
Replika Press Pvt. Ltd.
Plot No. A229, DSIDC Narela Indl. Park
Delhi 110 040

To

My wife,

NAHAR

and daughters

LOPA, SWATI AND TITHI

CONTENTS

LIST OF TABLES

LIST OF GRAPHS

ACKNOWLEDGEMENTS

RESEARCH WHICH has finally led to this work was begun way back in 1983 as a Humboldt Research Fellow at the South Asia Institute of Heidelberg University. During this long time (indeed much longer than anticipated) I have incurred obligations to many institutions and individuals. Among the former I am particularly indebted to Alexander Von Humboldt Foundation (of Germany) for awarding a Research Fellowship of two years duration. Major part of the primary research for this work was conducted during this period. My thanks are also due to the British Council for a short-term fellowship at the School of Oriental and African Studies (University of London), Indian Council of Historical Research for making it possible to work at Delhi and Calcutta, as well as to the Bangladesh University Grants Commission (especially its two successive Chairmen, Prof. M.A. Bari and Prof. Shamsul Hoque) for funding my trip to Lahore. The American Institute of Bangladesh Studies also deserves thanks for arranging my trip to the University of Pennsylvania (and several other universities in the USA) where I finalized my manuscript. I would also like to thank the authorities of Dhaka University for granting me leave of absence from time to time.

My work also owes a great deal to many individuals. Names of many of them will appear in the following pages, but some of them deserve special mention here. Prof. Dietmar Rothermund sponsored my Humboldt Fellowship, suggested the subject of research, provided me with detailed comments on my manuscript and arranged its publication. I wish to record my sincere gratitude to him for all his support and encouragement. I also acknowledge my debt of gratitude to Prof. K.N. Chaudhuri who has always been a source of inspiration. I am also indebted to Prof. A. Razzaque and Prof. M.R. Tarafdar for the keen interest they showed in my work. Dr. Clive Dewey, Prof. Imran Ali and Dr. Neeladri Bhattacharya shared with me their expertise and drew my attention to important source materials. Dr. Imran Ali (and some of his colleagues and students at the University of Management Sciences) also helped me in several ways during my stay in Lahore. Some of my colleagues in Dhaka University helped me in some way or the other. Those among them who deserve special mentioning are Prof. A.M. Chowdhury, Prof. Monwaruddin Ahmed, Dr. Nurur Rahman Khan, Dr. Sushil Ranjan Howlader, Dr. Khan Abdul Matin, Dr. Ratan Lal Chakravorty and Dr. Khairul Hossain. I sincerely thank them all. Thanks are also due to Prof. S. Bhattacharya, Dr. Inayat Allah Baluch, Dr. David Ludden, Dr. T.R. Sareen, Dr. Himadri Banerjee, Dr. Rajat K. Roy, Dr. Tajul Islam Hashmi, Dr. Suranjan Das, Shahnurur Rahman and Dr. Raman Mahadevan for all the encouragement that I received from them. I also thank Nurul Islam, Mainuddin and Moihammed Ali for their secretarial assistance. I am also indebted to Humayun Kabir, Akijuddin Howlader and Mir Nazrul Islam for their help in the finalization of the manuscript.

Finally, I acknowledge the generous co-operation received from the staff members at India Office Library and Records, National Archives or India (Delhi), National Library (Calcutta), Punjab Secretariat Archives (Lahore), Heidelberg University Computer Center and libraries of the School of Oriental and African Studies, London School of Economics, South Asia Institute (Heidelberg), Jawaharlal Nehru University, University of Pennsylvania and Dhaka University.

I am, however, entirely responsible for all errors of logic or facts.

M. MUFAKHARUL ISLAM

INTRODUCTION

The agricultural sector of the Indian subcontinent experienced a market expansion under the British colonial rule. Total volume (and value) of agricultural production increased through the extension of cultivation and, to some extent, improvement of the yield rates of certain cash crops. What is more, as India changed her role as the exporter of industrial goods to that of agricultural produce, and the domestic market also expanded with the development of certain modern industries and urban centres, much greater emphasis came to be laid on market-oriented production. The two interrelated trends were characteristic of all parts of the subcontinent, but several considerations suggests that these were one of the most pronounced in the Punjab province. Thus, Punjab achieved a higher rate of increase in the field of both cash and food crop production.[1] Moreover, whereas at the all-India level food production began to decline early in the twentieth century, in the Punjab the positive trend continued through to the end of the colonial era. Yet again, Punjab was a major supplier of agricultural goods to the outside world and other parts of the subcontinent. The impact of the widening of market was more pronounced in the Punjab even though this province did not come under British rule till the middle of the nineteenth century.

This relative dynamism of Punjab agriculture was a function of several favourable factors—construction of roads and railways, introduction of new varieties of seeds, restoration of peaceful conditions after years of anarchy and, most important of all, massive expansion of irrigation facilities, particularly in the public sector. Indeed, given the magnitude of progress achieved in this field it will be safe to argue that the dynamism of Punjab agriculture during the foreign rule was mainly due to the development of irrigation infrastructure. Immediately after annexation the British government repaired, renovated and expanded the old canals and then, beginning from the 1880s, built a number of new water works with the help of borrowed funds. Incidentally, Punjab received the lion's share of the government's investment in canal-building activities.[2] At the same time the methods of lifting water from wells improved significantly and this led to an expansion of well-lands. The result was that by the close of the British rule Punjab came to possess one of the largest irrigation system in the world, irrigated area exceeding 50 per cent of the total cultivated area. Indeed the colonial government played such an entrepreneurial role in constructing and maintaing a vast network of canals and distributaries as well as in the field of well-irrigation that it may well be described, in relation to the Punjab province, as an Hydraulic Regime.

Motives behind Canal Construction Programme

Why did the British government launch such a canal (and railways) construction programme in an era when otherwise (especially in the field of industrial development) it was committed to *laissez faire* doctrine? According to Paustian the explanation is simple enough—the government wanted to prevent the outbreak of famines by augmenting food production.[3] The same explanation has been suggested by several other scholars—Gustafson,[4] Bhatia[5] and Tomlinson.[6] But it is too narrow to be taken seriously. Firstly, it is true that following the recommendations of the Famine Commission (1880) certain canals were constructed for famine relief, but progress in this area was very slow. Thus, in 1900-1 preventive works accounted for only seven per cent of the total capital outlay and three per cent of the irrigated area.[7] The Indian Irrigation Commission (1903) laid down a criterion which was expected to facilitate expansion of preventive works, especially in areas like the Deccan, the Central Province and Bundelkhand where canals were urgently needed.[8] However, by 1919-20 these works could increase their share of the total capital outlay in the irrigation sector only upto eight per cent.[9] Secondly, the Punjab was not one of the famine-prone provinces in the subcontinent, but she received the bulk of the borrowed fund earmarked by the government for canal-building activities. Overall, as aptly pointed out by Shigemochi, if canals had really been constructed for the prevention of famines:[10]

> We would have difficulty in explaining why investment in irrigation work was concentrated in a least susceptible famine region such as the Punjab and why the Protective or famine relief works were given low priority in terms of capital outlay and rate of expansion.

Paustian refers to two other objectives mentioned by the contemporary officials as causal factors behind canal construction programme—reduction of population pressure in densely populated districts and creation of "villages of a type superior in comfort and civilization to anything which had previously existed in the Punjab".[11] But these explanations cannot be taken seriously either. There are many areas in the subcontinent with much higher population density, but it is well-known that the government did not contemplate any corrective steps. Actually, we would argue that reduction of population pressure in certain parts of the province was a necessity for the success of the canal construction programme. For most of the area commanded by the canals had little settled population and consequently the governemnt had to launch a colonization programme. With regard to the other objective, it is true that while alloting land in the colonies the government stipulated certain conditions for the creation of villages of a superior type. But it will be equally unconvincing to suggest that such an objective was a factor behind government decision to borrow fund in the London money market to construct canals in the Punjab.

In his pioneering work in Punjab canal colonies Imran Ali has suggested that the government sought to achieve a wide range of objectives through canal construciton and colonization programme; it wanted to promote agricultural development, expand the revenue base, enlist political support and facilitate recruitment in the army.[12] Imran Ali identifies the motives correctly, but this explanation is not adequate in the sense that it does not distinguish between the primary and secondary motives, or between ends which the government wanted to achieve from the canals and those that could be pursued because the canals brought vast lands under cultivation. For example, it is true that in allocating large chunks of colony lands to the landlords, retired army personnel and civilian officials as well as war veterans the government was motivated by political and military considerations. Military considerations are also apparent in the case of camel and horse-breeding grants. Similarly, government decision to allocate the highest proportion of colony lands to the peasant grantees was partly due to the fact that the Punjab peasantry provided the largest number of recruits for the army. Indeed, as additional lands became available for distribution the government were in a position to pursue objectives which were not directly economic in nature. But these were not the primary or even the main considerations behind the government initiative in canal construction. Actually we would suggest that for a fuller understanding behind government intervention in this area we need to view it in its wider perspective, the new pattern of relationship that was evolving between India and England in the second half of the nineteenth century. The subject is familiar enough, but given its importance in the present context a brief digression will be in order.

Need for expanding market

The British economy was undergoing a rapid transformation from the beginning of the Industrial Revolution in the middle of eighteenth century. In the changed circumstances Britain needed (a) expanding market for the products of her mills and factories and (b) supply of agricultural raw materials and wage goods. Consequently, with the growing strength of the industrial sector in the national economy the importance of Indian possessions came to be viewed in a different light. Hitherto, during the regime of the East India Company India was considered important as a source of mercantile profit and tributes, but now during the ascendancy of the industrial burgeoisie India was expected to play the role of an economic satellite, a supplier of raw materials for, and purchaser of the products of, the metropolitan industries. As aptly pointed out by one commentator, "Englishmen, inspite of holding contrary opinions about the way India should be linked to Britain, nevertheless expected that India could emerge as an excellent source of raw materials and a neverfailing market for the finished products of English factories. In all seriousness they considered this to be the ideal economic policy to be pursued in India."[13] Opportunities for the employment of British citizens and investment of capital as also the use of the Indian army for imperial purposes were, no doubt, important attraction of the Indian empire, but it would be safe to assert that the primary justification of the *Pax Britannica* lay in the need for converting India into a satellite of the British economy.

Two developments of the second half of the nineteenth century further reinforced this need for assigning a complementary role to the Indian economy. Firstly, during the 1870s British industries began increasingly to face competition from the rapidly industrialising countries of America and Europe and in the succeeding decades Britain ceased to be the prime industrial nation. Moreover, the intensification of

German and French expansion in South-East Asia, Oceania and Africa enhanced India's importance for the development of British economy. Secondly, the Civil War in the United States reduced the export of American cotton to the European market and this immediately increased the demand for Indian cotton.

The two roles (supply of primary produce and import of manufactured goods) which thus came to be assigned to India were interrelated, given that both these objectives could be fulfilled through the development of India's productive powers in the agricultural sector. For while the development of agriculture was necessary for the increased supply of raw materials and wage goods, such exports could enable India to pay for her imports from Britain (and earn a surplus). This explains why it became the "confirmed conviction"[14] of the rising class of British industrialists, merchants and bankers and their spokemen within and outside the Parliament that India should specialize in agricultural production and the Indian government should go beyond the negative policy of removing all obstacles to commercial intercourse, and promote larger production and export of agricultural raw materials through the construction of ports, harbours, roads, railways and irrigation works. Incidentally construction of public works for augmenting cotton production was advocated even by John Bright who, otherwise, repeatedly emphasized the purely moral duty of Britain to India. The same was true of James Wilson who belonged to the Cobden School of Anti-imperialism. For one of the five tasks which he set himself, on becoming the first finance member of the Viceroy's Council, was the promotion of "public works and roads with a view to increase production of cotton, flax, wood and European raw materials".[15]

Opinion differed as to whether incentive should be provided from the demand side (construction of railways) or through the supply-oriented policy of undertaking irrigation projects. One school of thought led by Sir Arthur Cotton maintained that the limited resources available with the government should be concentrated on the construction of irrigation-cum-canal projects. India possessed considerable water resources and labour and if the available water was stored properly the problem caused by the periodic failures of rains could be solved and even deserts be brought under the plough. Moreover, canals could be used for water transport. To defend his contention Sir Arthur also referred to the high rates of returns on the investment in the Kaveri, Godavari and Krishna irrigation projects. On the other hand, there was a strong body of opinion which criticized the protagonists of canals and favoured the construction of a railway network. They advanced several arguments against large-scale investment in water works: irrigation was not the sole remedy for low agricultural productivity, export crops like cotton and sugarcane did not need an assured water supply, all rivers were not equally perennial, return on investment would not be as high as in Madras, etc.[16] The Indian nationalist spokesmen were of the opinion that "it was of greater importance to construct irrigation works than more railways" for while railways could only lead to a more equitable distribution of available foodgrains in the different parts of the country, "irrigation works could prevent famines and scarcity by augmenting production".[17] But despite their opposition and the pontification of Lord Lytton that arguments about canals and railway network "are no less out of place than arguments as to the relative value of food and air",[18] priority was given to railway construction.[19] R.C. Dutta alleged that in doing so the British made "a geographical mistake. They needed few canals in their own country, and they therefore, neglected canals in India."[20] But as a matter of fact the explanation lies in the sustained agitation of the Lancashire manufacturers and British planters in India in favour of greater investment in railway construction.

Thus, as the Industrial Revolution progressed, there was a growing realization that steps should be taken to promote larger production and export of agricultural goods from India. As pointed out by Marx in his despatches to the *New York Daily Tribune,* during the regime of the East India Company the ruling oligarchy had no interest in developing India and they ignored the fact that agriculture in India was "not capable ofbeing conducted in the British principle of *laissez faire* and *laissez aller*".[21] However, the more the industrial interests became dependent on Indian market, the more it felt the necessity of creating fresh productive powers in India. Till recently, Marx continued, the ruling classes of Britain had "but an accidental, transitory and exceptional interest in the progress of India. The aristocracy wanted to conquer it, the moneyocracy to plunder and the milliocracy to undersell it." Now the new industrial and manufacturing classes "discovered that the transformation of India into a reproductive country has became of vital importance to them" and accordingly they felt that they should "gift her with means of irrigation and means of communication".[22] "You cannot continue to inundate a country with your manufactures", the British industrialists realized, "unless you enable it to give some production in return."[23] Incidentally, Marx also pointed out how irrigation works were paying rich dividends: "irrigated lands, even in districts near the Ghats, pay three times as much in taxes, afford ten to twelve times as much employment, and yield twelve or fifteen times as much profit, as the same area without irrigation".[24]

The demand of the British industrial interests to assign a special role to India as the buyer of manufactured goods and supplier of primary ones was often couched in language which stressed mutual benefits to be derived by both the countries. In this view there was no conflict of interest between colonies specialising in agricultural production and England producing industrial goods; their interests were reciprocal. "We have machinery and steam power for weaving, India has the cultivators and land for growing cotton without limits" said an East India merchant and added: "Why should our looms ever stand idle or the Indian expose his bare back to the sun?"[25] The same sentiment was expressed even more categorically by the policy makers. Britain was transformed into a major industrial power because "by her very nature and by her talents" she was destined to be so. Similarly economic development of India would take care of itself, it was confidently argued, if she could develop those resources (food and raw materials) for which she had a natural advantage. Indian nationalist spokesmen as well as some dissenters in England were quick to point out the defects of specialization in agriculture, and they advocated state intervention for the simultaneous development of agriculture and industry. But the British policy-makers remained committed to *laissez faire* doctrine in the field of industrial development.[26]

Role of political economy

The notion of complementary development within the empire and the stress on India's comparative advantage in agricultural production received powerful support from the views of the classical and neoclassical economists or the so-called principles of Political Economy.[27] According to these economists, by widening the extent of market and scope for the division of labour, international specialization, permits greater use of machinery, stimulates innovations, overcomes technical indivisibilities, raises the productivity of labour and enables the trading countries to enjoy increasing returns and achieve economic development. As pointed out by Meier, these dynamic aspects of the international division of labour were not central to classical and neo-classical thoughts, but there was nonetheless some recognition, particularly in classical theory, of the dynamic and growth-transmitting aspects of trade above and beyond the static gains arising out of comparative advantage. In this interpretation gains from trade were entirely consistent with gains from growth and, as a matter of fact, the latter could be expected to increase *pari passu* with the extension of foreign trade. J.S. Mill was the foremost champion of this view. According to the theory of comparative advantage, he said, trade leads to a "more efficient employment of the productive resources of the world" and this may be regarded as the "direct economical advantage of foreign trade. But", he continued, "there are, besides, indirect effects which must be counted as benefits of high order." One of the most significant of these indirect effects was the tendency of every extension of the market to improve the process of production. "A country which produces for a larger market than its own, can introduce a more extended division of a labour, can make greater use of machinery and is more likely to make inventions and improvement in the process of production."[28] The case of for India's specialization in agricultural production continued to be defended in the subsequent period. Thus, as late as in 1911 the young Keynes criticized the "current opinion in India, where a considerable part of the educated classes seems to desire with patriotic fervour the industrialization of their country". He added: "In my opinion such a change is not, in the future which one can foresee, either desirable or likely. It is an unfortunate consequence of the British connection that industrialism should present itself to the Indian as the royal road to prosperity and to a dignified position among nations."[29]

The same Political Economy which suggested that specialization in the production of primary goods could be the basis of substained economic development in India also advocated a positive role by the colonial government in the construction of certain public works—railways, roads, ports, harbours and canals. It is true that according to the classical economists economic development was a process initiated by the dynamism of individual entrepreneurs, but it was equally a part of the classical theory of economic policy that public works had to be undertaken by the government, because these provided the necessary framework "within which private economic activity could be successfully carried on".[30] Thus, Adam Smith believed that the state had three functions, one of which was the construction and maintenance of certain public works as these could never be the interest of any individual or group of individuals to construct and maintain them. Vigorous state role in certain specific areas was also advocated by J. S. Mill, Bentham, McCulloch and Senior. Overall, the principles of Political Economy could be invoked to justify not only India's specialization in the production of primary goods but also certain degree of government intervention to promote such specialization.[31]

Expansion of tax base

It is then clear that the primary objective behind canal (and railways) construction programme was to deepen and strengthen the colonial relationship by promoting the larger production and export of agricultural raw materials and wage goods from India. But the government did not need only real (exportable) surplus to enhance India's foreign exchange earnings so that she could pay for the imports and the invisible items in her balance of payments with Britain (and thus play a vital role in financing British deficits with Europe and America). At the same time the government had to obtain a financial surplus[32] to maintain the army (sometimes to be used outside India to serve British interests), the expensive and expanding civil and police administration, pay the Home Charges[33] and fulfil a host of other obligations. It is important to mention here that between 1877-8 and 1892-3 Home Charges increased by 62 per cent (from Rs 162 to Rs 262,000,000) and averaged Rs 194,000,000 per year (or 26 per cent of the Indian government's gross revenue).[34] But while revenue needs were increasing, income from customs dwindled from the mid-century through the 1890s with the movement for free trade. Land revenue did inccease,[35] but only from Rs 200 to Rs 250,000,000 during the period 1873-95. Further increase was not considered feasible, given the experience of the past years when revenue default and forced sales of land had become customary in different parts of the country. Incidentally, with the annexation of the Punjab (1849) the British had also reached the natural limits of colonial expansion in India. Further territorial conquest would have taken the British to Afghanistan and Nepal. But neither place offered much possibility of large land revenue. In the circumstances the government were obliged to explore other avenues for expanding the tax base. Thus, the second most important objective behind canal-building activities was to increase the financial resources of the government from direct and indirect sources-water rates, sale of colony lands and enhancement of land revenue. It is true that lands brought under cultivation by the canals sold at a nominal price and water-rate was fixed at a low level, thus leaving a considerable proportion of the enhanced value of land and out turn with the grantees. But financial consideration remained one of the most important objectives behind canal construction programme. In this connection the opinion of the Indian Irrigation deserves particular mentioning for it went to be extent of observing that productive (irrigation) works in the Punjab were "deliberately undertaken more or less as commercial undertaking".[36]

It is then evident that in launching the canal (and railways) construction programme the government was motivated, primarily by two considerations: (a) help India perform her special role as the supplier of agricultural produce and importer of industrial goods and (b) expand the revenue base. In other words, the government sought to extract both a real and financial surplus from India's agriculture. It is also obvious that these two objectives were interrelated inasmuch as both could be achieved by increasing the volume of marketed surplus. Thus, the argument can be further simplified by saying that the main impetus to cannal construction was provided by the need for promoting market-oriented production in the agricultural sector, especially the production of those crops (cotton and wheat) which commanded a ready market in the outside world. Incidentally this explains why commercialization of agriculture was considered so important that for the colonial government it virtually meant agricultural development. It is true that at the all-India level the process of commercialization did not go for enough and, notwithstanding the near-universal recognition of the importance of public works, overall progress in this field remained limited. But the basic concern behind the attempt to promote "agricultural development" in India is not in doubt. Thus, to conclude this part of the discussion, it would not be tenable to argue that political/military considerations, or the desire to reduce population density in certain areas or create villages of superior type provided the primary incentive behind government intervention in the construction of irrigation works in the Punjab, this intervention has to be understood in the context of the evolving economic relationship between India and England.

It is not difficult to explain why Punjab received the bulk of the attention of the colonial government in the field of canal construction. About 80 per cent of the total area of the province lay within the gently sloping alluvial plains and five rivers (the Indus, Jhelum, Chenab, Ravi and Sutlej) cut them into a series of low parallel watersheds called *Doab*. These plains were fertile, but rainfall was inadequate, uncertain and unevenly distributed over different seasons. The result was that at the time of annexation cultivation was virtually restricted to the districts along the foothills of the Himalayas, to the riverains and areas where well-irrigation could be practised. The uplands beyond these areas presented a desolate look, hardly supporting any settled population. The government realized that these tracts (which it took over and called Crown Waste Lands) could be turned into prosperous agricultural areas if perennial canals were constructed and a process of colonization started. The task was very difficult but the prospects appeared to be bright. It was also realized that if canals were built in these areas, it would be possible to avoid the problems and costs which arose in areas (for example, in United Province) having settled population and/or practising irrigation from other sources. Secondly, it was found that the rivers coursing these areas could be relied upon for unfailing supplies of water throughout the year. This advantage, coupled with the fact that the surface of the arid plains was even, made it possible to construct canals at a low cast. Finally, the canals constructed in the first phase proved remunerative in the highest degree and this itself gave a powerful stimulus to investment in this field. Thus, the fact that under foreign rule Punjab came to possess the largest irrigation system in the world was due to the combination of a number of favourable factors—topography, soil and climatic conditions, presence of vast wasteland, the snow-fed rivers and of the canals (or their ruins) constructed at earlier times which fired the imagination of the foreign rulers as well as profitability of the canals built in the first phase. As pointed out by the Indian Irrigation Commission, "In no other province are there greater facilities for extensive irrigation; in none is it more urgently necessary or more keenly appreciated."[37]

Need for detailed research

Contemporary observers, especially the officials of the colonial government, used a plethora of superlatives to describe the importance of the Punjab canals. Thorburn described them as "colossal, productive and life-creating and preserving".[38] The Abiana Committee took the view that canal water "was the greatest boon which the government can confer upon its subjects. . .".[39] One commentator was of the opinion that the "irrigation works have the same value to our great agricultural province as the coal and iron mines to Britain and gold mines to the Transval".[40] Some historians of the post-1947 period have described the importance of the canals in similar vein. Thus, Nijjar has suggested that the canals "revolutionized the economic position of the Punjab and its people".[41] Yet again, one economic historian has described the canals as among the "greatest monuments to Britain Rule".[42] But such descriptions have not been matched by detailed research into the precise nature of the impact of the canals on the agrarian economy and society of the province. Recently Imran Ali has dwelt at length on the whole range of considerations which influenced government policy with regard to the disposal of colony lands, extent of land received by different social classess and other related issues. His work constitutes a significant contribution to the extent that it goes, but, as aptly pointed out by himself, several other important aspects of the irrigation system still remain to be explored by the historians.

Incidentally we would argue that the whole subject of irrigation has remained under-researched. For till recently the only study of this topic constituted a single chapter of a book on the Uttar Pradesh agriculture in the closing decades of the ninteenth century.[43] In this connection it should also be mentioned that irrigation from private sources, especially wells, has hardly received any attention from the historians. For example, in Whitcombe's essay on "Irrigation" in the *Cambridge Economic History of India* there is virtually no reference to irrigation practized from sources other than public canals. This is so even though after nearly one century of canal building activities by the government in the Punjab province one in every six acres/received water supplies from the private sources.

There is another lacuna in Whitcombe's approach. For she confines her attention to the evolution of irrigation policy of the colonial government from the first quarter of the nineteenth century. As about the importance of canals in agricultural terms she believes that "this remains something of a mystery"[44] because information on the value of canals water in relation to the levels of production and productivity are not available. Results of the crop-cutting experiments are available, but, she continues, these furnish neither the amount of water supplied nor the number of waterings nor the sources of water. But we would argue that notwithstanding the limitations of the relevant data it is possible to make some idea about the importance of irrigation works in agricultural production and productivity. The kind of data Whitcombe has in mind are always hard to come by except from experimental stations and, when available, the interpretation of such data is normally beyond the competence of the historians. Secondly, it is misleading to give the impression that the value of the canals has to be judged in terms of their contribution to productivity alone. Productivity is, of course, one of the most important field of investigation, but this is by no means the only area of concern for the historians.

As a matter of fact irrigation system, particularly of the Punjab type, provides a large number of themes for historical research and there are several considerations which suggest that these themes deserve more attention than they have received so far. Firstly, large-scale public investment in canal building activities represents one of the best example of "colonialism of development" as distinguished from "colonialism in the cheap".[45] It would therefore be of particular interest to assess the impact of the irrigation system on the social economic and political life of the province as well as its (irrigation system) importance to the colonial state. Secondly, research into this area may throw light on the debate as to why agriculture in the two parts of the Punjab province has shown greater dynamism in terms of technological progress and enhanced productivity in the post-1947 period.[46] Needless to mention, the study should also shed light on the many questions that have exercised the mind of the scholars generally about Indian agriculture and agriculturists under British colonial rule. Finally, a vast range of quantitative and qualitative data is available on different aspects of the working of the irrigation system and other related subjects—land grant, sale and mortgage of land, acreage and production of major crops grown in irrigated and non-irrigated lands, prices of agricultural goods and their export, occupational distribution of population, extent and varieties of tenancy, level of rent, distribution of agricultural land among different classes of rural people, problem of debt, moneylenders and moneylending etcetera.[47] These data are of such a nature that it is possible to undertake in-depth research on specific themes for the province as a whole or the socio-economic life of a single village,[48] or the working of a particular canal. But the present study had only a limited objective in the sense that it seeks to provide a general description of the irrigation system and an introductory analysis of some of the more important aspects of its impact on the agrarian economy and society.[49]

The study begins with an account, in broad, descriptive terms of certain select aspects of the irrigation system: an historical outline of the private and public canals, different methods of lifting water from wells and the technological progress achieved in this field, land grant and water management in the colonies. The second chapter presents the annual rates of change in the total irrigated area and the area irrigated from different sources of water supply. In this connection we try to explain the preference of the cultivators for one type of water supply over another and indicate the sources of expansion of irrigated acreage. This is followed by an attempt, in the third chapter, to discern the impact of an assured water supply on the crop patterns. We use certain farm-level data to see whether the crops that increased their share of the total irrigated area were profitable ones. An important development in the province during the period under review was a marked increase in sharecropping tenancy. In the fourth chapter we focus on the impact of irrigation system as it evolved in Punjab on the incidence of tenancy. The fifth chapter deals with the impact of irrigation on land transfer, a subject which was much discussed in the Punjab official and political circles. This is followed by a discussion of the financial results of public investment in canal construction in the sixth chapter. The sources and nature of the statistical data used in throwing light on these aspects of the agrarian economy and society of the province as well as the financial returns for the government are mentioned in the respective chapters. Finally, we present as extended summary of the main findings of the study in the seventh chapter.

NOTES

1. For details see George Blyn, *Agricultural Trends in India,* 1891-1947; *Output, Availability and Productivity* (Philadelphia, 1966), pp. 99, 119. Elsewhere I have suggested that grwoth rate in the Punjab was actually lower than estimated by Blyn. But still it seems that Punjab agriculture fared better than the agriculture of any other province. For the latter view see M.M. Islam, "Trends in Crop Production in the Undivided Punjab: A Reassessment" in Clive J. Dewey (ed.), *Arrested Development in India: The Historical Dimension* (Delhi, 1988).
2. By the late 1930s cropped areas (in million acres) irrigated by government canals were as follow: Punjab 12.3; Madras 7.2; U.P. 5.3; Sind 4.7; Bihar 0.7; N.W.F.P. 0.5; Orissa 0.4 and Bombay 0.4. For details see V. Anstey, *The Economic Development of India,* 4th edn (London, 1957), p. 616. Simultaneously steps were being taken by the government for the construction of railways and metalled roads. In 1919-20 there were 1441 miles of railway and 2938 miles of metalled roads in the Province. For details see H. Calvert, *The Wealth and Welfare of the Punjab* (Lahore, 1922), Ch. IV.
3. Paul W. Paustian, *Canal Irrigation in the Punjab* (New York, 1925), pp. 23-5.
4. W.E. Gustafson and R.B. Redlinger, "Delivery of Canal Water in North India and West Pakistan", *Economic and Political Weekly,* Vol. VI, No. 52 (Dec. 25, 1971).
5. B.M. Bhatia, *Famines in India: A Study in Some Aspects of the Economic History of India 1860-1905,* 2nd edn (Bombay, 1965).
6. B.R. Tomlinson, "The Indian Agricultural Problem, 1900-1950: Land, Labour and Capital", paper presented at the 12th European Conference on Modern South Asian Studies (Berlin, 1992). Tomlinson mentions that this was the objective in the nineteenth century.
7. Hirashima Shigemochi, *The Structure of Disparity in a Developing Agriculture: A Case Study of the Pakistan Punjab* (Tokyo, 1978), p. 30.
8. *Report of the Indian Irrigation Commission,* 1901-3 (Calcutta, 1903), p. 112.
9. Shigemochi, *op. cit.,* p. 30
10. *Ibid.,* p. 23.
11. Paustian, p. 48. He also mentions that another objective was to expand the revenue base. As elaborated later this was indeed a primary objective, but Paustian seems to underestimate its importance by referring to unimportant or less important factors.
12. Imran Ali, *The Punjab Under Imperialism* (New Delhi, 1989), Ch. 2.
13. S. Ambirajan, *Classical Political Economy and British Policy in India* (Cambridge, 1978), p. 53.
14. Cited by F.F. Clairmonte, *Economic Liberalism and Underdevelopment* (Bombay, 1960), p. 16.
15. Cited by Ambirajan, pp. 53-4.
16. For details see *ibid.,* p. 54.
17. For the views of the nationalist spokesmen see Bipan Chandra, *The Rise and Growth of Economic Nationalism in India* (New Delhi, 1966), pp. 207-14.
18. Cited by Ambirajan, p. 54.
19. The total capital expenditure by the government on major irrigation works up to 1902-3 amounted to nearly four crores of rupees, but the total outlay on state and guaranteed railways

amounted to 359 crores by 1904-5. For details see Bipan Chandra, p. 208. For the relative share of these two sectors in the subsequent period see M.J.K. Thavaraj, "Capital Formation in the Public Sector in India: A Historical Survey 1898-1938" in V.K.R.V. Rao *et al.* (eds.), *Papers on National Income and Allied Topics* (Bombay, 1960).

20. R.C. Dutt, *The Economic History of India in the Victorian Age* (London, 1903), p. 367.
21. Marx's despatch to the *New York Daily Tribune* (henceforth *NYDT*), 25 June 1853.
22. Marx's despatch to *NYDT*, 8 August 1853.
23. Marx's despatch to *NYDT*, 11 July 1953.
24. Marx's despatch to *NYDT*, 8 August 1853.
25. Cited by Ambirajan, p. 219.
26. The author of the Report of the Royal Commission on Industrialization in India (1916-18) opined, "Throughout the nineteenth century the policy of the Government was controlled by the doctrine of laissez faire in commercial and industrial matters...".
27. This part of the discussion is based on Ambirajan, Ch. 2; G.M. Meier, *The International Economics of Development* (Tokyo, 1968), Ch. 8 and H. Myint, "The Classical Theory of International Trade and the Underdeveloped Countries", *Economic Journal*, LXXIII (1958).
28. Cited by G.M. Meier and R.E. Baldwin, *Economic Development: Theory, History, Policy* (New York, 1966), p. 43.
29. Cited by Clairmonte, p. 127.
30. R.D.C. Black, *Economic Thought and the Irish Question* (Cambridge, 1959), p. 162.
31. For details see L. Robbins, *The Theory of Economic Policy* (London, 1952), Lecture II.
32. The concepts of physical and financial surplus are borrowed from T.J. Byres who uses these in a different centext, for details see his, "Land Reform, Industrialization and the Marketed Surplus in India: An Essay on the Power of the Rural Bias" in David Lehman (ed.), *Agrarian Reform and Agrarian Reformism* (London, 1974).
33. Though the Home Charges were ultimately remitted in the form of goods, in the first place, these were paid out the revenue of the Indian government. For details on the mode of this payment see A.K. Benerji, *Aspects of Indo-British Economic Relations*, 1858-1898 (New Delhi, 1952).
34. John Adams and Robert Craig West, "Money Prices and Economic Development of India 1861-95", *Journal of Economic History*, XXXIV (1979).
35. Dietmar Rothermund, *An Economic History of India* (New York, 1988), p. 43.
36. *Report of the Indian Irrigation Commission*, 1901-3, Part 1 (London, 1903).
37. *Indian Irrigation Commission*, Part 2, p. 1.
38. S.S. Thorburn, *The Punjab in Peace and War* (Lahore), p. 280.
39. *Report of the Abiana Committee* (Lahore, 1934), p. 41.
40. K.S. Ahmed, "The Agricultural Geography of the Punjab" (unpublished Ph.D. Thesis, London, 1939), p. 168.
41. B.S. Nijjar, *Punjab Under British Rule* (1848-1947), Vol. 3 (New Delhi, 1974), p. 30.
42. W.J. Macpherson, "Economic Development in India under the British Crown, 1858-1957" in A.J. Youngson (ed.), *Economic Development in the Long Run* (London, 1972), pp. 144-5. According to the view the other monuments were railways, law and the civil service.
43. E. Whitcombe, *Agrarian Conditions in Northern India*, Vol. 1, *The United Province Under British Rule*, 1860-1900 (London, 1972). For a critique of Whitcombe's study see Ian Stone, *op. cit.*
44. *Ibid.*, *Cambridge Economic History of India*, Vol. 2, p. 728.
45. Richard G. Fox, *Lions of the Punjab* (University of California Press, 1985), p. 62.
46. Xinru Liu, "Small Land-holdings in the Punjab: The Historical Perspective", *Punjab Past and Present*, 16-18 (32).
47. These data are hardly mentioned in W.E. Gustafson and K.W. Jones, *Sources on Punjab History* (Delhi, 1975).
48. T.K. Kessinger, *Vilayatpur*, 1848-1968 (Berkeley, 1974).
49. The study starts from 1887-8 because it is from this year that statistics on irrigated area sown with the main crops were made available in the government publications.

Chapter 1

THE IRRIGATION SYSTEM

IRRIGATION HAD been extensively practised from very early times in the territories which constituted the province of Punjab during the British colonial rule.[1] This is clear from the epigraphical and literary evidence relating to the ancient and medieval periods,[2] extensive network of canals or their ruins[3] encountered by the British as well as the existence of 137.6 thousand wells in the province in the middle of the nineteenth century.[4] However, the vast and modern irrigation system which is possessed by the two parts of the province (particularly the Pakistani one) and which has given them a unique position in the irrigation map of the world was developed during the British colonial rule. This chapter is an attempt to present, in broad and descriptive terms, the outlines of this system—a resume of the canal building activities by the colonial government, land grant and water management in the canal colonies, and other sources of artificial water supply, especially wells.

1.1 Government Canals

Such canals as existed in the province at the time of annexation were mostly of the inundation type. Generally speaking, these canals along the Jamuna, Indus and the Lower Chenab rivers obtained[5] supply only when the parent streams were in flood and the adequacy or otherwise of this supply and consequently the irrigable area was wholly dependent upon seasonal conditions. The only perennial canal, i.e. canal which had at its head a weir or barrage across the bed of the parent stream, was the Hasli canal. Originally constructed in 1613 by Emperor Shahjahan's engineer Ali Mardan Khan to bring water from the Ravi to Lahore, it was extended by the Sikhs to Amritsar to fill the sacred tank about the Golden Temple. Anyway, in some parts of the province these inundation canals were in serviceable conditions but in others these had become useless. The necessity of constructing new water-works was realized quite early by the British, but, to begin with, the government turned its attention towards the improvement of the existing works, rather than the construction of new ones.

The first major work taken up after annexation was that of Bari Doab Canal. The work began with the intention of remodelling the Hasli canal, but later developed into a much larger work. The upper Bari Doab Canal, as it was later called, started irrigating in 1861. The total mileage in operation in 1945-6 included 360 miles of main canal and branches and 1479 miles of distributaries, involving a capital outlay of Rs 221,000,000 (Table 1.1). The same year the canal irrigated 1.5 million acres in Lahore, Gurdaspur and Amritsar. Although later irrigation projects were undertaken

TABLE 1.1: STATISTICS ABOUT THE PUNJAB CANALS (1945-6)

Canals in Operation	*Mileage*	*Year of Completion*	*Capital Outlay ('000 Rs)*	*Area Irrigated ('000 acres)*
Productive				
Western Jamuna Canal	2,120	1886	20,435	1,018
Western Jamuna Extension	286	1895	14,350	121
Sirhind Canal	2,000	1886-7	26,692	2,312
Upper Bari Doab Canal	1,839	1878-9	22,125	1,485
Lower Bari Doab Canal	1,499	1917	23,963	1,540
Upper Chenab Canal	1,399	1917	43,311	830
Lower Chenab Canal	2,904	1899-00	15,165	2,921
Upper Jhelum Canal	715	1917	47,157	379
Lower Jhelum Canal	1,193	1917	26,729	1,005
Sutlej Valley Project	3,526	1933	88,788	2,235
Haveli Project	1,582	1939	37,840	1,080
Total Productive	19,053		409,871	14,926
Unproductive				
Indus Inundation Canal	779	1849-50	3,496	299
Shahpur Inundation Canal	214	1870-1	232	69
Muzaffargarh Inundation Canal	767	1896	1,737	366
Gaggar Canal (British)	78	1898-9	388	27
Western Jamuna Canal	180	1886-7	1,266	-
Gagger Canal	53	1898-9	291	-
Sirhind Canal	1,973	1886-7	14,435	-
Sutlej Valley Project	-	1933	59,222	-
Thall Project	-	-	40,207	-
Total Unproductive	1,833		53,261	762

Source: Statement II-C in *Administration Report of the Punjab Public Works Department (Irrigation Branch)*, 1945-6.

to derive economic advantages by bringing new lands under cultivation, contemporary evidence make it clear that the primary motive behind this canal was to give employment to disbanded Sikh army. The canal was completed in 1878-9.

As mentioned earlier, one system of inundation canals found by the British included the series of canals along the Jamuna river and the oldest of these canals in the Punjab was the Western Jamuna Canal. First built by Sultan Feroz Shah in the fourteenth century to supply water to the royal lodge at Hissar, the canal was renovated during the reigns of Mughal Emperor Akbar and Shahjahan. Except for a small part which continued to be locally worked, this canal had fallen into disuse in the middle of the eighteenth century. But the British were able to use many of these canals as part of the renovated Western Jamuna Canal. This canal was completed in 1886. The Sirsa branch was added in 1889-95. Over 200 miles of main canals and distributaries were in operation in 1945-6 and the total capital outlay amounted to nearly Rs 35,000,000 (Table 1.1). The same year the canal irrigated more than one million acres in parts of Ambala, Karnal, Rohtak and Delhi in British territories.

In 1868 another large irrigation project was launched namely the Sirhind Canal. This canal was planned to take off from the Sutlej river at Rupar and irrigate the British districts of Ludhiana, Ferozepore, Hissar and the native states of Patiala, Nabha and Jind. Under an agreement signed in 1873 one-third of the cost of this canal was paid by the rulers of these native states and, in return, they were entitled to an equal proportion of the available water supply. These rulers agreed to undertake the maintenance and management of their own distributaries. Two factors led to a scarcity of labour in the construction of this canal. Firstly, the area through which the canal was being constructed was waste and far away from the settled villages. Secondly, these villages were affected by the famines of 1860 and 1868. The government solved this problem by employing prison labour on excavation work.

The Sirhind Canal was the first of the British canals in the Punjab built under the policy of financing irrigation projects from loans raised in the London money market. This policy was first proposed by Lord Dalhousie.[6] But instead of adopting this proposal the government had decided to promote irrigation projects by the same indirect agency as that by which they were extending railways—by guaranteeing certain return to private companies. Soon the government became disillusioned about the prospect of constructing and maintaining canals under private initiative, and in 1866, the Secretary of State approved of the scheme for raising loans in London specially for irrigation works, provided that such works were expected to be remunerative or productive. Basically, this test required that only such projects as would, after ten years of their opening, yield an annual income equal to the interest on the capital expended on their construction should be accepted. From 1868 irrigation works constructed with the help of borrowed funds were described as Extraordinary Public Works. At the same time the administrative framework was restructured. In 1866 the Public Works Department was divided into military, civil (including irrigation) and railway branches and the next year saw the creation of the provincial Public Works Department. Though the Government of India and the Secretary of State in London exercised, through the Inspector-General of Irrigation and the Public Works Secretariat in Calcutta, powers of superintendence, direction and control, the execution and management of irrigation projects now became a provincial function. This arrangement continued till 1919.

The Hasli Canal, Western Jamuna Canal and the Upper Bari Doab Canal were built through the populated parts of the province. The Sirhind Canal also supported some tracts with settled agriculture. Thus, on the whole, these four canals were constructed with a view to improving existing agricultural system. But beyond these territories were the vast tracts which were practically deserts supporting no settled population. The rainfall in these areas was less than 10 inches per year and the level of sub-soil water was as low as 50 to 80 feet, while water raised from that depth was unsuitable for irrigation purposes. The most important phase in the history of Punjab irrigation started when the government took over these unclaimed lands as Crown Waste and embarked upon an ambitious scheme of canal construction and colonization with the help of borrowed funds. Thus, the government assumed a two-fold responsibility—construction of canals and the newer and more difficult task of devising colonization scheme. This element of colonization distinguished canal construction in the Punjab from other parts of India. In the latter, canals were built to improve existing agricultural system. But since there was no or little resident population in the Crown Waste Lands it was necessary, simultaneously with the construction of canals, to physically transfer the whole communities into the area thus opened up. Between the years 1885 and 1940 nine canal colonies were developed in West Punjab. These were situated in the inter-fluvial tracts or Doabs west of the Beas-Sutlej and east of the Jhelum rivers. These tracts were known as Bari Doab, Rechna Doab and Jech Doab and these lay between the Beas-Sutlej and the Ravi rivers, the Ravi and the Chenab rivers and the Chenab and Jhelum rivers. The canal colonies were irrigated by perennial canals that led off from a river spanning weir and their network of branches and distributaries.[7]

The first canal to involve colonization was the Lower Sohag-Para Canal taken out of the Sutlej some 45 miles below Ferozepore. For about 19 miles of its main line the canal was a development of the natural Sohag Nala after which it was purley an artificial channel—bifurcated into Sohag and Para branches. This was the only inundation canal of some importance which was built entirely from imperial funds. It had two head channels—Tali Bagar head and the Lalu Gudar head. This peculiarity was due to the sudden change in the river in 1890 when the original head at Lalu Bagar silted up leading to a complete failure of water supply. The alternative head at Lalu Gudar was created to guard against the recurrence of this danger. The gross area commanded by this canal was 242.2 thousand acres out of which 86.3 thousand acres were government land. In 1925 the Sohag-Para Canal was incorporated into the Sutlej Valley Project.

The next canal to be opened for irrigaton was the Sidhnai Canal. The first proposal to irrigate land in the Multan district by means of a canal drawing its supply from the Sidhnai reach of the Ravi was submitted as early as in 1856, but it was not till 1882 when the Punjab government was asked by the Indian government to submit proposals for new irrigation works to be constructed out of loan funds that the Sidhnai project was reconsidered. A fresh estimate was submitted in 1883 and it was accepted the same year. The area commanded by the canal was 4.2 million acres of which 2.3 million acres were government waste lands available for colonization. The main canal was opend in 1886 and the colonization process was completed in less than three years. The project was extended during the course of construction as it was realized that there was a danger of the riverain villages below the weir being affected by the diversion of water into Sidhnai. To gurad against this danger it was decided to construct three small canals—the Koranga, Fazal Shah and Abdul Hakim. These small canals, hardly bigger than major distributaries, were opened for irrigation in 1891. In one respect the Sidhnai canal was a peculiar one. For although potentially a perennial work (as it was provided with permanent headworks), the canal was used only for *Kharif* harvest. This was so because the available water supply of the Ravi during the winter season was drawn off by the Upper Bari Doab Canal at upper reaches (at Madhopur). Thus, Sidhnai Canal held a position midway between a perennial and an inundation canal.

The next year (1887) saw the opening of the Lower Chenab Canal. The need for artificial irrigation in the Rechna Doab was mooted way back in 1862 and several proposals were considered at different points of time thereafter. But once again it was not till 1882 when the Indian government called for a statement of projects likely to prove remunerative that definite steps were taken. The provincial government proposed the construction of an inundation canal, but it was to differ from similar canals in as much as it envisaged the irrigation of the high plateau of the Doab and not the lower riverain tracts to which the inundation canals were normally confined. The scheme was sanctioned by the Secretary of State in 1884 and the canal was opened for irrigation in 1887. But it was a complete failure. Firstly, the estimated cost was exceeded and therefore the possibility of earning a profit was considered remote. Secondly, during the flood season the channels silted heavily and, as the river fell to its normal level during the winter, it was found that the silted channels were too high to permit the normal flow from the river. In the circumstances colonization became an impossible task and accordingly in the same year (1887) a revised estimate was submitted together with a note by the Superintending Engineer Major J. Ottley in which he proposed the conversion of the Chenab Canal from an inundation to a perennial canal. This proposal was finally accepted by the Secretary of State in January 1890. But shortly afterwards, the Superintending Engineer Major R.E. Jacob proposed the enlargement of the Rakh or Midala Branch and the construction of two other branches—the northern or Lodhran (afterwards designated the Jhang) and the Southern or Sheikhpura (afterwards called the Gugera). These proposals were designed to distribute all the available water of the Chenab river over the whole area of the Doab which could be commanded. He also suggested that the construction of the projected Jhelum Canal should be postponed till the Chenab Canal had been completed. These proposals were accepted in July 1892. Thus, the Lower Chenab Canal was opened as an inundation but completed as a perennial canal. With nearly three thousand miles of main canal, branches and distributaries and involving a capital expenditure of Rs 50,000,000 this was the second largest of the canals constructed by the colonial government. This canal created the Lyallpur Colony and irrigated parts of Gujranwala, Sheikhpura and Jhang districts.

The Lower Jhelum Canal was opened in 1901. The idea of irrigating the Jech Bar drawing water supply from the Jhelum goes back to 1847-8 when Sir Herbert Edwardes mentioned that, from the experience of his journey in this area, he was sure about the irrigability of this tract. He said, "a more level tract could not be found and I venture to say that no obstacle could be met to a canal from Jalalpur on the Jhelum to the junction of that river with Chenab".[8] A specific proposal for the irrigaton of this tract was submitted in 1876, but the Punjab government accepted the opinion of the Financial Commissioner Sir R. Egerton that any scheme in this area must be postponed till the Rechna Doab Canal had been constructed. Another proposal was submitted to the local government in 1882. A detailed survey of the tract was carried out and its results were embodied in a *Report on the Jhelum Canal Project*. Submitted by Major R.E. Ottley in 1887 this report recommended that canal should be constructed with its off-take from the Jhelum river at Rasul. This project was accepted by the Secretary of State in 1888, but as per the suggestion of Sir R. Egerton the construction was postponed till the completion of Lower Chenab Canal. The excavation of the main line of the canal started in 1897 when the famine-stricken people of the Hissar district were readily available for employment. Meanwhile the scope of the canal was further widened to embrace an area of 1,140,000 acres (subsequently increased to 1,531,000 acres). Crown Waste Lands constituted about 50 per cent of the culturable area commanded by this canal. Colonization started in 1902 and by 1921 more then four thousand acres were allotted.

Canal-building activities in the Punjab received a powerful stimulus from the recommendation of the Indian Irrigation Commission (1901-3). The Commission was of the opinion that railway construction (along with the construction of protective canals) recommended by the Famine Commission (1880) as a measure for famine protection had played its part in the field of famine insurance and it was now time to develop food resources. With regard to the Punjab, the Commission pointed out that "it is here that the greatest progress has been made in irrigation works of this (perennial) class; and in spite of this fact it is here that there is still the greatest field for their further expansion".[9] The Commission was prompted to take this view because it found that the Punjab canals yielded the highest rate of return and the utility of artificial water supply was appreciated most by the cultivators of this province.

The most important irrigation work constructed in the province following the recommendation of the Irrigation Commission was the Tripple Canal Project. Described by one author as a "striking monument"[10] to the farsightedness of two officers who originally proposed the scheme (Col. S.L. Jacob, the designer of the Lower Chenab Canal, and Sir James Wilson, the Settlement Commissioner) and the engineers responsible for its construction the Tripple Canal was so called because it linked the waters of three rivers (Jhelum, Chenab and the Ravi) and made it possible to construct three canals—Upper Chenab, Upper Jhelum and Lower Bari Doab Canal. Till 1901 all proposals for the irrigation of the lower portion of the Bari Doab (between the Beas and the Ravi rivers) were based on the assumption that water supply must be drawn from the Sutlej river, given that whole of the winter volume of the Ravi was already hypothecated to the Upper Bari Doab Canal. Actually a detailed scheme was drawn up in 1901 on this assumption. But this scheme was opposed by Colonel Jacob and Sir James Wilson. Their contention was that the waters of Jhelum were much greater than could be utilized in the tracts between the Jhelum and the Chenab and irrigation of the Lower Bari Doab represented the last possibility of their beneficial use. They further contended that the Sutlej water would be needed for the further development of irrigation facilities on its two sides. Ultimately their opinion prevailed and the Tripple Canal Project was designed by the Chief Engineer, Sir John Benton. The project transferred the surplus water of the Jhelum into the Lower Bari Doab and thus made it possible to plan the Sutlej Valley Canal Project which, in latter years, was to extend cultivation in the deserts of Rajputana.

To transfer the water of the Jhelum a regulator was constructed at Mangla wherefrom the Upper Jhelum Canal carried the Jhelum water into the Chenab, discharging it above the headworks of the Lower Chenab Canal at Khanki. The latter river was thus fed by the Jhelum water and the Chenab water so freed was taken from a network (at Marala) into the Upper Chenab Canal. This was the second link of the

Tripple Canal Project. Running southwards, it crossed the Ravi at Balloki below which it was known as Lower Bari Doab Canal. The Upper Jhelum Canal was opened in 1915-16 and completed in 1917. It irrigated the Upper Jech Doab. The Upper Chenab Canal was the largest perennial canal in the world and it irrigated the Upper Rechna Doab. The canal was opened in 1913-14 and completed in 1917. The Lower Bari Doab Canal which irrigated portions of Lahore, Montgomery and Multan districts was opened in 1913. Together these three canals commanded 3.3 million acres of culturable area in 1941-2 and involved a capital outlay of Rs 114,000,000. These canals converted the most desolate portion of the province into one of the most prosperous agricultural areas.

Irrigation became a (Reserved) Provincial Subject under the Indian Council Act, 1919. As a result the provincial government had now increased financial resources at its disposal and was, therefore, in a position to take greater initiative in constructing water works. At about the same time the old system of classifying irrigation works was abolished. Till 1921 government canals were divided into three categories: Productive (or Major), Protective and Minor. As mentioned earlier, Productive canals were constructed from funds raised by loans, and expected to earn a net profit within ten years of their completion. The Protective canals, first recommended by the Famine Commission (1880) were built out of current revenue for protection against famines. The Minor works were also financed from current revenue and these included works not classed as Productive or Protective. A distinction was made between Minor works for which capital and revenue accounts were maintained and those for which such accounts were not maintained. In 1921, it was decided that all works of public utility were to be classed as either "Productive" or "Unproductive" according to their financial results, without reference to the source of funds. The "Productive" works were those which yielded sufficient revenue within 10 years to cover their working expenses and the interest charges on their capital cost calculated at certain stipulated rates.[11] Moreover, any work of public utility whether directly "Productive" or not could be financed by borrowed funds.

Two important projects were executed by the government after the implementation of the Council Act—the Sutlej Valley Project and the Haveli Canals. The Sutlej Valley Project, made possible by the Tripple Canal Project, had three purposes. Firstly, it was designed to control the volume of water entering the series of inundations canals from the Sutlej by the provision of weirs and head regulators. Secondly, its purpose was to expand the area irrigated by the existing canals by including the low-lying areas of the river valley in the command areas. Finally, it aimed at providing perennial water supply to the uncultivated uplands on both sides of the river. The project was sanctioned by the Secretary of State in 1921 and construction work was immediately started on four canals—Pakpattan, Dipalpur, Eastern and Mailsi. The whole project was completed in 1933. The Haveli Project, first planned in 1917, was designed to divert the water of Jhelum and Chenab into Ravi near Sidhnai by constructing a barrage at Trimmu. The Canal was opened in 1939. During 1939-40, the Karam, Ganesh and Talini Canals of the Muzaffargarh canal system as well as Sidhnai Canal and the Indus Inundation Canal were amalgamated with this system.

1.2 Disposal of Colony Lands

The waste lands (or the Crown Waste, as it was called) made cultivable by the construction of canals was disposed of by the government (a) through grants to different social groups under a great variety of names and (b) auction sales.[12] The unit of allotment varied between 22.5 and 27.5 acres in size depending upon the size of the rectangles and squares in the particular colony. The grantees may be said to have been of two broad categories: those with service conditions and those without any condition. These two categories may again be divided into three groups: (i) peasant grantees, (ii) *nazrana*-paying grantees and (iii) military grantees.

The peasant grantees constituted the largest single group among the recipients of colony and the major portion of land grant was made to them. They included both the immigrants from other parts of the province and indigenous population of the irrigated tracts. The immigrants were selected chiefly from the Jullunder, Hoshiarpur, Lahore, Gurdaspur and Amritsar districts from among the landowners or members of the "agriculture castes". The Jats, Arains, Kambohs, Sainis and Gurjaras were the most important caste groups among the peasant grantees. For the sake of homogeneity and social harmony grantees of the same caste were settled in the same village.

The *nazrana*-paying grantees included (i) the Yeoman and Capitalist grantees in the Chenab Colony, (ii) Civil grantees in the Jhelum and Chunian Colonies, (iii) grants to the landed gentry in Lower Bari Doab Canal and (iv) grants to individuals in return for services rendered to the government. The most important recipient of these grants were the large and small landlords and members of the urban bourgeoisie—merchants, contractors, lawyers, doctors and retired public servants. In caste terms the latter groups belonged, by and large, to the "non-agricultural castes" as defined under the Act of 1901. They came from such castes as Khatris, Aroras, Khojas, Banias and Sheikhs. They were called *nazrana*-paying grantees because they paid an annual fee to the government as a sign of the ownerwhip of land which had previously been state property. Sometime it took the form of rent while at other times it was simply an additional tax payment.

Military grantees received allotments in recognition of their military service to the state. In most cases these grants were of the same size and subject to the same conditions as the peasant grants in matters relating to payment of *abiana* (water rate), government dues, cultivation of soil and period of probation. The official justification for the military grants was that though soldiers belonged to the same social groups as the civil grantees, they had little chance of selection in the civilian allotment owing to their absence from the village.

Grants with service conditions were made in the Lower Chenab, Lower Jhelum and the Lower Bari Doab Canal. Among the service conditions stipulated by the government the most important were the maintenance of camels for transport and mares for breeding purposes, planting of tree nurseries and production of quality seeds. The demand of the Indian Army for horse and mules provided a steady market for the animals raised by the grantees. With regard to the production of seeds, in the Lyallpur Colony the government let out an area of 500 acres for seed farming on half produce basis. The subtenants were required to sow selected seeds according to the Agricultural Departments' directives and half of these seeds obtained by the Department as rent was distributed among the agriculturists. Again, in the Lower Bari Doab Canal two "seed farm" grants were made. The two grantees—Sir Jogindra Singh and H.T. Conville—had to devote stipulated acreages for the production of seed which was then procured by the government and other buyers. Mr. Conville's farm proved more successful. It did not only supply seeds but crop experiments were regularly conducted on it and farm machinery was used.

In all the colonies a certain portion of the irrigable land was sold to men of substance through public auction. These sales were more profitable than grants because land could be sold at market prices. But it was only in the Nili Bar Colony that area sold through auction assumed some significance. Thus, while auction sales accounted for only 12 per cent and 8 per cent of the total land disposed of by the government respectively in the Chunian and Lower Bari Doab Canal in the Nili Bar Colony nearly half (45 per cent) of the land was reserved for sale. The decision to sell such a large portion of the irrigable land in this colony was prompted by the financial stringency of the Punjab government in the post-war years. Except in the Nili Bar Colony sales through public auction were mostly confined to the town sites.

We propose to analyse the implications of the land grant policy of the government in a separate chapter. However, at this stage it may be added that the role of the government in the colonization programme went far beyond the selection of colonists and disposal of the new landed resources. For under the Government Tenants (Punjab) Act, 1893, it imposed several conditions on the peasant grantees in matters relating to sanitation, succession, residence and right of alienation. Residence of the grantees in their respective colonies was insisted upon because it was feared that otherwise they would convert themselves into rentiers. The regulation of succession was considered necessary to avoid rapid sub-division and fragmentation of holdings. Lastly, no transfer by sale, mortgage or gift was allowed except a sublease for one year in the case of tenants-at-will and seven years in the case of occupancy tenants. It was further provided that once the tenant-grantees had paid their dues to the government and fulfilled all other conditions of tenure they would be entitled to receive a *Sanad* conferring on them the right of occupancy in land after a period of five years.

But as time went on many defects of the Act of 1893 became apparent. Most important of all, it was realized that the officials should have been empowered to enforce conditions of tenure by summary or executive process, i.e. without recourse to civil courts. Accordingly, in 1906 the Punjab government introduced in the Legislative Council a Bill for the Colonization of Government Lands to supercede the Act of 1893. The Bill was passed by the Council but vetoed by the Governor-General because, meanwhile a great agitation had flared up in the colonies against the bill.[13] The government took this agitation seriously and a high-powered committee known as the the Colonies Committee was appointed to review the whole question. The Committee recommended major concessions to the colonists[14] and these were, to a large extent accepted by the Punjab government and incorporated into the Colonization of Government Lands (Punjab) Act, 1912. It was now provided that (a) the present procedure as regard grant of lease and grant of occupancy right would continue, (b) proprietary rights would be granted 5 years after the grant of occupancy rights, (c) customary law of succession were to prevail in case of grantees who had acquired proprietary rights, (d) residence in the colonies would be compulsory in the initial period when the grantees were government tenants, but this restriction would cease to operate with acquisition of proprietary rights and (e) in the case of those who had become proprietors there would be no restriction on the right of alienation except those provided by the Act of 1901. Thus, the Punjab government abandoned the powers which it had assumed to introduce institutional innovations in the canal colonies.

1.3 Water Management

The headworks of the canals, main lines, branches and distributaries were constructed and maintained by the Irrigation Department, but the field channels or water-courses by means of which water was finally supplied on to the fields were usually constructed and invariably maintained by the irrigators themselves. Usually a separate water-course was given for an area of 30 to 70 rectangles or squares (or 750 to 1750 acres). The efficiency of water-course primarily depended on its correct alignment and this, in turn, was an important factor behind the "duty" or efficiency achieved by a particular canal system. Therefore, considerable care and technical skill were needed for the selection of the correct site. When a water-course was constructed by the government, it became a property of the irrigators and the cost was recovered from them in instalments. A water-course could belong to only one irrigator or it could be shared by two or more cultivators. If, in the latter case, the irrigators failed to agree to an equitable distribution of water among them, the matter was referred to the officials of the Irrigation Department, canal water was carried on to the fields in two methods: the flow system in which water was delivered at a level not lower than that of the land and the lift system in which case water was delivered from a lower level and had to be raised to the lands.

The area irrigated by a water-course or outlet was called an "Irrigation *chak*".[15] The method followed was to turn the full discharge of the outlet on the each square or rectangle of the *chak* in succession and then, within each square or rectangle, on each acre (*killa*) in succession. This method was followed because in the winter water supply was insufficient to keep the canal running continuously and, therefore, the distributaries had to be closed on a system of rotational turns called *waribandi*. For this purpose the day was divided into eight *pahras* of there hours each and the practice was to allow each holding a turn (*wari*) of two to four *pahras*. To facilitate distribution and economise water use each acre or *killa* was required to be divided into a number of compartments or *kiaris* (eight, according to the regulation of the Irrigation Department). The irrigators were expected to arrange matters in such a manner that they had, for each compartment, water supply for 20 to 30 minutes. But the irrigators were not favourably disposed to this rule requiring the division of a field into *kiaris* or small beds and in some cases the rule remained a "dead letter".[16] Their objection to this system was due to fact that it took much longer to water a given area and this was a weightly consideration, given that water was available only by turns. Secondly, if *kiaris* were insisted upon in the case of preliminary watering, they had to be broken up for the subsequent ploughing and then made again after sowing, thus involving additional labour. The *waribandi* system could not also be strictly enforced, for in many cases the cultivators in the canal colonies were regularly taking water from a water channel out of turn. They resorted to this practice known as *warashiknis* in different ways, the most important of which was making an opening in a channel bank by means of a cut or a hole. An ordinary cut of a important distributary could affect 30 to 40 villages and cause the entire channel to be closed for repairs. In 1929 it was estimated that a cut on such a distributary could mean a loss of Rs 100,000 to the affected cultivators and Rs 10,000 to the government. Most of the distributaries suffered 10 to 15 per cent per annum, involving an annual loss of Rs 10,00,000 for the irrigators and Rs 100,000 of the government. *Warashiknis* was most common during the sowing period and/or when water was short in supply. Officials of the Irrigation Department were aware of the *warashiknis*, but they could not do much to prevent it for a variety of reasons. Firstly, *waribandi* was not legally binding. Secondly, there were many corrupt officials. Thirdly, colonists who resorted to such practices were so powerful that petty officials did not dare antagonize them.[17]

The Irrigation Department laid down, for every canal, what was known as "designed permissible area" of irrigation. It represented that proportion of gross commanded and culturable area which was meant to be irrigated every year. Several considerations led the government to stipulate the permissible area. Firstly, it was felt that the colonists should be encouraged, on ground of hygiene, to maintain the waste spaces created in the colonies. Secondly, cultivators outside canal colonies practising irrigation from wells and other sources should not be tempted to the colonies by easy access to canal water. Finally, and most importantly, the fear of waterlogging dictated that irrigated area should not exceed a certain limit. Accordingly, the Irrigation Department laid down the permissible area which should be irrigated by each of the canals and their branches in the dry and cold season. For example, in the case of the *Bar* lands of the Lower Jhelum Canal the permissible limit was 75 per cent of the commanded area and the ratio of *rabi* to the *kharif* was two to one.[18] In other words, if the commanded area was 10,000 acres, 750 acres were to be irrigated annually, 500 acres in the dry and 250 acres in the summer season. A further rule provided that 88 acres of *kharif* crops and double this area of *rabi* crops should be irrigated for each cusec of water supplied from an outlet. Allowing for 10 per cent absorption in the distributary this meant that at the distributary head one cusec was needed for 80 acres and at the canal head (allowing 18 per cent absorption in the main canal and branches) one cusec for 65 acres (or 130 acres in the *rabi*). Thus, for irrigating 250 acres in the *kharif* season it was necessary to provide at the outlet for the delivery of (250 ÷ 88) 2.84 cusecs of water. The "duty" of water (or the area or crop which could be matured by each cubic foot of water (per second) in the *rabi* season was meant to be double because during this season the channels ware in full supply approximately half the time of the *kharif* season.[19]

But, as the Colony Manual pointed out, the provision of permissible area was a misnomer in as much as it was never the intention of the government to enforce this limit rigorously.[20] It is also clear that the irrigators did not feel constrained by this provision either. For, at the aggregate level as well as in the case of every canal the area actually irrigated exceeded the permissible area almost every year.[21] Thus, the extent of area irrigated did not depend on this provision *per se*, but on the availability of water supply and other relevant considerations. But yet this provision relating to permissible area created a controversy. For while the cultivators insisted on calling it *haqq abphasi* because it implied guarantee, the government argued that it was never their intention to give such a guarantee and accordingly, it called the permissible area *reqba adphasi mujiawiza*. What seems to have prompted the irrigators to take such a position is the fear that water might be short in supply and even the permissible area would not be cultivated. For, as pointed out by the concerned officials, despite continued efforts to equalize supply some villages at the tail of long distributaries were in a "chronic.state of uncertainity and sometimes of actual distress".[22] This was due partly to ordinary technical difficulties and partly to the "greed of villagers higher up which at critical times steal the water intended for their less favourably situated neighbours".[23] In this connection it may be recalled that in many districts, especially the canal colonies, the cultivators solely depended on the canals for their water supply.

1.4 Private canals

Besides the state-owned perennial and inundation canals there were several private canals irrigating an area of 495,000 acres in 1943-4. Such canals existed from very early times and when the British conquered the province some of those canals were taken over by the government at the owners' request. At the same time the government encouraged the construction of private canals by leasing out waste lands. This programme helped extend cultivation in several districts, particularly in Shahpur. All except one of the private canals in this district were owned by the members of the Noon and Tiwana families. Of the private canals in other districts the most important were the Grey Canal in Ferozepore, Sarusti Canal in Karnal, Gharah and Gharkhana Canals in Montgomery and the Shahan Nahar Canal in Hoshiarpur district. The operation of these canals gave rise to certain problems, because these canals did not irrigate the lands of their owners only, the surplus water was sold to other cultivators. There were complications even when the canal was constructed solely to irrigate the lands belonging to the owners. For state interest was involved in the withdrawal of water from the rivers and the supply channel could not be constructed without affecting lands owned by other cultivators. Moreover, water charge could be collected by the canal-owners at high rates, thus interfering with government claim to land revenue. The North Indian Canal and Drainage Act (1873) which was drawn up to deal with large canals did not make rules ragarding these problems. Accordingly, the government passed the Punjab Minor Canals Act which legislated for smaller irrigation works, whether private or public.

1.5 Wells and Other Sources of Water Supply

At the time of the annexation of the province and for many years thereafter wells constituted the most important source of artificial water supply in the province.[24] The suitability of a tract for successful well-irrigation primarily depended on three factors:

(a) depth of sub-soil water (b) constancy of supply and (c) quality of the water.[25] The first condition was the most important, for as a rule, irrigation by wells was not practised where the depth of water was below 35 feet, for the cost of lifting water from beyond that depth was excessive compared to the value of crop grown.[26] The *khadir* areas, that is, the areas adjacent to the rivers and the sub-montane districts (Multan, Hoshiarpur, Dera Gazi Khan, Mazaffargarh and Gurdaspur) fulfilled these conditions more than the other tracts and accordingly these districts had the largest concentration of wells. In some parts of the predominantly well-irrigated (*chahi*) tracts of the province wells remained out of the use during the summer months when the areas they served were submerged by river floods. On the other hand, in some cases when the floods were good and the winter rains unusually plentiful and timely the wells remained unused throughout the year or a single well watered an area in which, if the season was bad, three or four wells were brought into operation.[27] In these areas wells constituted a form of insurance. Where the wells were unaided either by flood or canal, the system of well-irrigation had several peculiarities. Each portion of the field was moistened, ploughed and sown in turn. But when the wells were aided, the whole area had to be simultaneously ploughed and sown with crops before the effects of the autumn drenching had passed off. The unaided wells were generally deeper than others and these required a superior breed of cattle to operate them. On the other hand, the area attached to these wells were larger in size with a higher yield per acre.[28] Wells were also sunk in the predominantly canal-irrigated tracts (called *nahri* areas), but the primary purpose was to use these wells as an insurance against drought, and these were put into operation only during dry season.[29] Indeed, in the case of the Sidhnai and Sohag Para Colonies, which depended on inundation canals, construction of irrigation wells was stipulated by the Irrigation Department as one of the conditions on which lands were to be held. Moreover, indirect incentive was provided for the construction of wells in certain portions of the Upper Chenab Colony which received irrigation only during the *kharif* season.[30]

Wells were generally of three categories: (a) those lined with brickwork and called *pakka* or masonry (b) without brickwork or called *kachcha* (these being lined with nothing more than grass) and (c) semi-masonry, i.e. masonry in one stratum only. An average *pakka* well served a larger holding than a *kachcha* one. Moreover, while a *kachcha* well lasted only two or three years and then fell in, a *pakka* well lasted much longer, from 30 to 100 years[31] in case there was no serious fall in the water level. But a *pakka* well was much more expensive than a non-masonry or semi-masonry well. For example, in the case of Lahore district[32] it was estimated that a *kachcha* well cost Rs 30 to Rs 40 to sink but a masonry well cost from Rs 500 to Rs 900. The area that a well could irrigate depended on a large number of factors: the nature of the soil, the character of the season, the quality of the cattle employed and the industry of the irrigators, depth of sub-soil water, crop composition, condition of the well and the method of lifting water. Given this situation, it is not surprising that the area normally irrigated with the help of one well varied from district to district. For example, in the Salt Range[33] one well normally irrigated 2.5 acres, but the average area commanded by a well was 28 acres in Gujranwalla[34] and 54 acres in the Ara Circle of Shahpur district.[35] The number of wells in the province increased from 137,000 in 1848-9 to 347,000 in 1943-4. The capital value of the wells at the cost of Rs 800 for each well may be said to have been of the order of Rs 278,000,000[36] in 1943-4 as against Rs 345,000 in the major or productive irrigation works.

In the case of jointly-owned wells[37] the arrangement for watering depended on the number of co-sharers. This means that the modes of dividing water varied with the variation in the number of shares. Each share was allotted a certain period called *wari* divided into *pahras* of three hours each. If there were only two or three shares in a well then the *wari* extended to eight *pahras* (i.e. 24 hours); if the number of share was four or more, the period allotted to the *wari* was 12 hours or four *pahras*. The *wari* of 12 hours was by far the most common. A *wari* was never less than a two *pahras* or six hours and never more than eight *pahras* or 24 hours. In some cases the timing of the *wari* had to be changed artificially but in others it changed automatically. For example, if a well was held by A, B and C and if each sharer worked the well for four *pahras* A's turn at the well if the previous one had lasted from midnight to midday on Monday, would be from midday to midnight on Tuesday. Similarly, if a turn of two *pahras* was allowed to each of the six co-shares, the timing of each turn changed in regular order. But if the turn was of two *pahras* for each of the four sharers in a well, then the change had to be made artificially. The change was made in such a way that it gave the two night turns to the proprietors who had earlier day turns.[38] *Waris* always corresponded to the proprietary shares in the wells or to the proprietary shares represented by the extent of land held by the tenants. A one-third sharer in a well was not entitled to an extra long *wari* because he possessed an extra pair of bullocks.

Various methods were used to lift water from wells: Persian wheel (*Rahat*), *Charsa*, tube-wells powered by oil engines or electricity, *Dhenkli* and *Jhatta* or swing basket.

From references in the *Beburnamah* it is clear that there had been considerable use of *Rahat* as early as in the early sixteenth century.[39] But the period under review witnessed marked improvement in the operational efficiency as well as increased application of the Persian Wheel for lifting water from wells. Indeed, the progress in the latter field was so rapid that early in the twentieth century Persian Wheel became the most widely used water-lift in the province.[40] In the absence of relevant quantitative data it is not possible to estimate the number of Persian Wheels in use or the proportion of area irrigated under this method. But the available evidence from the *District Gazetteers* and other relevant sources clearly point to this conclusion. The progress was even more remarkable in certain tracts of the province in the sense that here well-irrigation virtually came to mean irrigation with the help of *Rahat*. For example, in the case of Jhang district it was reported that Persian Wheel was "the universal method of lifting water from wells".[41] Similarly, the concerned officials of Multan claimed that the only form of well in the district was the Persian Wheel well.[42]

The Persian Wheel was an intricate piece of construction and in this respect it varied from district to district. But basically it consisted of a large drum (*bair*) over which passed an endless rope or iron-ladder (*mahl*) with miniature buckets attached to it at a distance of one or two feet. The ladder with the buckets reached below the surface of water in the well and the drum with the buckets was revolved with the help of a simple round-about gear worked by a pair of bullocks.[43] The buckets were either earthenware vessels which contained from one and a half to two seers of water or were made of iron capable of holding three to six seers of water.[44] Sometimes wells were built large enough to allow two Persian Wheels at the same time. Such a well was called a *Wan*. When the water was near the surface and supply good, such wells were quite common.[45] A Persian Wheel was called a *Jhallar* when it was used to lift water from *nala*, streams ponds or canals. The *Jhallar* were of several types. The *Oral* or *Bahar Bedi* was a small contrivance having only a few pots but of a larger size, it was worked by one bullock only. The *Tangan* or *Utangan* was used when water was at a medium distance. A *Jhung* was one in which there were two wheels plying into the same reservoir. On the canals *Jhallars* were used for two purposes: either to irrigate lands on to which canal water could not flow or to prolong the period of watering when lower supplies were received at the beginning or the end of the cropping season.[46]

The improvement in the operational efficiency of Persian Wheels, mentioned earlier on, was achieved mainly in four ways. Firstly, while in the pre-colonial period the Persian Wheel was made of wood, iron-made Persian Wheels became fairly common from the early years of the twentieth century. A study of the efficiency of the two types of machines indicated that on an average iron *Rahat* was 28 per cent more efficient than wooden Persian Wheels.[47] However, in many cases this change from the old to the new remained partial in the sense that Persian Wheels were made partly of wood and partly of iron. Secondly, in many cases aluminium was substituted for the earthenwares in buckets. Thirdly, the introduction of a cheap form of metal roller bearing led to a considerable reduction in friction, thus permiting simple automatic lubrication of the moving parts.[48] Finally, though Persian Wheels were normally worked by bullocks, in some cases the contrivance was operated by electricity or oil engines.[49]

A *Rahat* had several advantages over the indigenous water lifts—*Charsa* or *Dhenkli*. Firstly, for a Persian Wheel, four to six bullocks were required as against six to eight for a pulley well. Secondly, a child or an old woman was enough to drive the bullocks for a *Rahat*, but to operate a *Charsa* two, sometimes even three, able-bodied men were needed.[50] Thirdly, the *Rahat* was very flexible and adapted for different water depths. For example, an increase in the size and number of buckets led to an increase in draft and discharge. Similarly, increasing the size of the *bair* increased the velocity of the chain draft and output, while decrease in size had the reverse effect.[51] Fourthly, the Persian Wheel commanded a larger area. It is true that the *Rahat* was more expensive. This was particularly so when all the parts of the contrivance were made of iron. The operation of this water-lift was more expensive also because it was normally used in *Pakka* wells. But given the fact that agricultural income was increasing from the turn of the century partly through the rise in the price level and partly because of the improvement of the yield rates of certain crops, the richer section of the peasantry could now afford the investment for the construction and operation of Persian Wheels and, consequently, there was now a shift to this latter water-lift.

The indigenous *Charsa* was the more commonly used water-lift till the early twentieth century. But, as mentioned above, it was gradually replaced by the Persian Wheel and ultimately towards the end of the period under review this device became virtually confined to the tracts (i.e. the south-east of the province) where water level was rather low and the rainfall good so that only few waterings were enough to mature crops. The *Charsa* was not as intricate a construction as a Persian Wheel was. Basically it consisted of a leather bucket made of a whole bullock hide with the corners cut off and tied to an iron-ring nearly two feet in diameter. The bag, capable of holding 24 to 42 gallons of water, was pulled by a pair of oxen by means of rope passing over a pulley at the top of the well. As the bag reached the level of the discharging trough it was emptied by an attendant appointed for the purpose. The bag was then again thrown in the well and filled with water by the man stationed there and the next load was ready. The task of the oxen was made easier by the fact that these had to walk down an inclined plane while pulling the water bag, but had to climb up when these were without the load.[52] Sometimes two pairs of oxen were used in which case three men were employed.

The agricultural engineers of the province were engaged in the task of devising means of increasing the supply of water in the wells. One such device was the practice of boring a hole in the well and putting in of a plain or tube-well pipe in the bore to connect the ordinary wells with the substarranean supplies. This practice started in the second decade of the current century and it was undertaken by the Well Boring Section of the provincial Agricultural Department. During the three years ending 1943-4, 540 wells were annually bored in the province and 81 per cent of these bores proved successful.[53] This improvement was later taken a step further and a form of irrigation intermediate between perennial canal and a good well was developed. This was done by inserting a strainer tube-well, a device by which the water in the deep-seated layers of course and sand could be raised to the surface by a pump driven by an oil-engine. These installations were often 260 feet in depth and were capable of watering 200 to 400 acres. The most common tube strainer, basically consisted of a copper mesh supported on a strong iron frame. In some cases the earthenware strainers evolved by the Agricultural Engineer, Mr Miller-Brownlie, were being used.[54] Total number of tube-well operated in the province in the late 1930s exceeded ten thousands.[55] Tube-wells worked by oil-engines was most cost-effective,[56] but Punjab was considered to be unsuitable for the expansion of this mode of irrigation because (a) in many areas water was brackish and (b) in the majority of cases yield rates began to decline after a lapse of time.[57] Moreover, capital cost of such a tube-well was very high, estimated to be Rs 4786 as against Rs 800 in the case of a bullock-driven Persian Wheel.[58]

Dhenkli or counterpoise lift was normally used by the poorer section of the cultivators. It consisted to a bucket made of iron or leather suspended by a rope to the end of a pole. The pole was evenly balanced by a counter-weight on a stout upright support on which it worked in a see-saw fashion. This lift was used by manual labour and the counterpoise was so heavy that no great exertion was necessary to depress it when it was empty or raise the bucket when it was full with water.[59] The *Dhenkli* was used to irrigate holdings which were too small for economical use of bullocks and the sub-soil water was not very deep. This device was used to lift water from tanks or wells five or six feet deep. Another method of lifting water was the *Jhatta* or the swing basket. It consisted of a basket or shovel like scoop and was swung by two men standing on two sides holding the rope. This device was suitable only for low lifts.

Three other sources of artificial water supply in the province categorized in the official publications as "other sources" were embankments, springs and streams. Embankments of great continuity were found in the Gurgaon district.[60] Many of these embankments gradually silted and for many years the government remained indifferent to them. These were restored to working conditions in the early 1920s when they were placed under the Drainage Board. Irrigation by means of embankments was also resorted to in Mianwali district, in the Salt Range and the hilly tracts.[61] In the hills lands were irrigated by means of small channels called *Kuhls* through which the waters of the hill streams were led to and distributed to terraced fields. These *Kuhls* were virtually the only source of artificial water supply in Simla and Kangra districts.[62] Sometimes 15 to 20 channels were supplied from a single stream. The channels were constructed and maintained with considerable labour. Perennial and non-perennial springs were the principal source of irrigation in Jhelum and Dera Gazi Khan districts. Thus, these methods of irrigation were of some importance only in certain areas of the province. Consequently, the area irrigated from embankments springs and streams constituted only a small portion of the total area. For example, in 1901-2 the "other sources" accounted for only 5 per cent of the area irrigated from all the sources taken together.

Thus, under British colonial rule the Punjab province witnessed a vast expansion of her irrigation infrastructure through the repair and renovation of the old inundation canals, the construction of a number of perennial canals with the help of borrowed funds and improvement in the method of lifting water from wells. The construction of the new canals in the Crown Waste Lands involved large-scale colonization and led the government to attempt a kind of "social engineering" in the canal colonies by imposing on the grantees certain conditions about inheritance, residence, land transfer, water management and sanitation.

NOTES

1. It may be argued that Marx exaggerated the importance of (a) irrigation in Indian subcontinent as a whole in the pre-colonial period and (b) the role of the government in the construction and maintenance of water work, but as far as the Punjab is concerned artificial water supply occupied an important place in the agrarian life of the province. For critiques of (a) and (b) see, respectively Angus Maddison, *Class Structure and Economic Growth: India and Pakistan Since the Moghuls* (London, 1971) and Harbans Mukhia, "Marx on Pre-colonial India: An Evaluation" in Diptendra Banerjee (ed.), *Marxian Theory and the Third World* (New Delhi, 1985).
2. Nand Kishore Kumar, "Hydraulic Agriculture in Peninsular India (*c.* 300 B.C. to 1300 A.D.)", *Proceedings of the Indian History Congress*, 1979 (Delhi, 1980), p. 5.
3. For details see *Punjab Gazetteer*, Vol. 1, 1908.
4. G.S. Chhabra, *Social and Economic History of the Punjab* (Jullunder City, 1962), p. 173.
5. Much of the description about the working of the canals and other sources of irrigation remains valid even today. But I have found it more convenient to describe everything connected with the irrigation system in the past tense. The discussion of the renovation and repair of the older canals is based on P.W. Paustian, *Canal Irrigation in the Punjab* (New York, 1925), Ch. II.
6. For details see Ian Stone, *Canal Irrigation in British India: Perspectives on Technical Change in a Peasant Economy* (Cambridge University Press, 1984), pp. 18-27.

7. The resume about the canal-building activities of the British government is based on Paustian, pp. 48-63; H.K. Trevaskis, *The Punjab of To-day: An Economic Survey of Punjab in Recent Years,* 1890-1925 (Lahore, 1931), Vol. 1, pp. 241-90; Fareeha Zafar, "The Impact of Canal Construction on the Rural Structures of the Punjab: The Canal Colony Districts, 1880 to 1940" (unpublished Ph.D. Thesis, University of London, 1981), Ch. III; K.S. Ahmad, "The Agricultural Geography of Punjab" (unpublished Ph.D. Thesis, University of London, 1939), Ch. IV; L. French, *The Punjab Colony Manual* (Lahore, 1931), Vol.1; Imran Ali, *The Punjab Under Imperialism,* 1885-1947 (New Delhi, 1989), Ch. 2 and Y.B. Mathur, *British Administration of Punjab,* 1849-75 (Delhi, 1966), Ch. I.
8. *The Punjab Colony Manual,* Vol. II, 1922 edn, p. 13
9. *Report of the Indian Irrigation Commission,* 1901-3, Part II (Provincial), Punjab, p. 1.
10. Trevaskis, p. 262.
11. For details see B.S. Saini, *The Social & Economic History of the Punjab,* 1901-1939 (Delhi, 1975), p. 207.
12. Disposal of Colony lands is discussed in great detail in *Punjab Colony Manual,* 1922 edn, Ch. II; Paustian, pp. 62-74 and Imran Ali, Chs. 3, 4.
13. N.G. Barrier, "The Punjab Disturbances of 1907: The Response of the British Government in India to Agrarian Unrest", *Modern Asian Studies,* Vol. 1, No. 4; Shyamala Bhatia, *Social Change and Politics in Punjab 1898-1910* (New Delhi, 1987), Chs. 5, 6.
14. For details see *Report of the Punjab Colonies Committee,* 1907-8.
15. For details see *Punjab Colony Manual,* Vol. 1, 1922 edn, Ch. 9; Trevaskis, pp. 292-4 and William Roberts and S.B.S. Kartar Singh, *A Text-Book of Punjab Agriculture* (Lahore, 1947), pp. 165-8.
16. *District Gazetteer* (henceforth *DG) Hissar,* 1915 (Lahore, 1916), p. 146.
17. Imran Ali, pp. 182-5.
18. *Punjab Colony Manual*, pp. 173-5.
19. This particular example is taken from the summary of statement submitted to the Royal Commission on Agriculture by the Chief Engineer of Irrigation Works (Punjab). The impression that is given in this statement is that the permissible limit for all the canals and all the canal-irrigated areas was more or less the same. But the *Colony Manual* (pp. 156-68) mentions a whole range of limits from 30 to 80 per cent.
20. *Colony Manual,* pp. 73-4.
21. For details see the relevant data made available in the *Punjab Public Works Department—Irrigation Branch: Annual Administrative Report with Statistical Statements and Accounts.*
22. *Final Settlement Report of the Jhang and Gugera Branch Circles* of the *Lyallpur District* (Lahore, 1925), p. 5.
23. *Ibid.*, p. 5.
24. Even as late as the quinquennium 1887-96 well irrigation accounted for 50 per cent of the total cropped area in the province.
25. *DG Multan,* 1901-2 (Lahore, 1902), p. 195. The District Gazetteers provide detailed information about the construction and operation of irrigation wells.
26. Roberts and Singh, p. 142.
27. *DG Muzaffargarh,* 1929 (Lahore, 1930), Part 1, p. 115. This means that an increase in the number of wells does not necessarily mean an expansion of well-lands.
28. *DG Multan,* p. 147.
29. *DG Ferozepore,* 1915 (Lahore, 1916), Part I, p. 189.
30. *Colony Manual,* p. 176.
31. *DG Gurgaon,* 1910, p. 118 and *DG Gujranwalla,* p. 120.
32. *DG Lahore,* 1916, p. 124
33. Saini, p. 205.
34. *DG Gujranwalla,* 1935, p. 120.
35. *DG Shahpur,* 1917, pp. 2-3.
36. According to Roberts and Singh's estimate the amount was Rs 250,000,000.
37. From the evidence available from the District Gazetteers it is not possible to say whether the jointly owned or the individually owned well was the dominant pattern. For, while the *DG*

Karnal (1918, p. 109) pointed out that wells were seldom the property of single person, *DG Sialkot* (1920, p. 103) pointed out that the old joint-stock wells had largely given place to individually owned wells.

38. This example is taken from *DG Jhang*, 1929, p. 114, see also *DG Gujranwalla*, p. 121.
39. Chetan Singh, "Well-Irrigation Methods in Medieval Punjab: The Persian Wheel Reconsidered", *Indian Economic and Social History Review*, XXII (January-March 1985).
40. Roberts and Singh writing early in the 1940s pointed out that some 20 years ago *Charsa* was very common, but then it was largely replaced by the Persian Wheel (p. 145).
41. *DG Jhang*, p. 111.
42. *DG Multan*, p. 184. In Amritsar district *Charsa* was not known. *DG Amritsar*, 1914, p. 4.
43. Trevaskis, p. 235.
44. Roberts and Singh, p. 143.
45. *DG Montgomery*, pp. 183-4.
46. *DG Multan*, pp. 153-4.
47. Roberts and Singh, p. 143.
48. Trevaskis, p. 236.
49. Roberts and Singh, p. 144.
50. *DG Ferozepore*, 1915, p. 190. The easy manner in which a Persian Wheel could be he operated was expressed by the saying "one eye is enough for a *Rahat*", *DG Karnal*, 1918, p. 108.
51. Roberts and Singh, p. 145.
52. For details see Trevaskis, p. 236 and Roberts and Singh, pp. 114-15. A modification of the *Charsa* devised by the Dept. of Agriculture eliminated the necessity of a man to empty the bag at the well head. But this form was too expensive to become popular. For details see the *Report of the Royal Commission on Agriculture in India*, Vol. VIII: *Evidence Taken in the Punjab* (London, 1927), p. 169.
53. Roberts and Singh, p. 146. The Government played a role in the expansion of irrigation by *Rahat* and *Charsa* in two other ways: (a) grant of *taccavi* loan and (b) concessions as regards revenue assessment. When a masonry well was constructed the land watered by it was exempted from that part of land revenue which was due to the existence of the well for 20 years from the start of operation. But the period was shorter when the well was constructed to irrigate land already assessed at water rates. When an existing well was repaired for the purpose of irrigation a similar concession was given for a maximum of ten years. No concession was given for unlined wells which cost little and were not permanent. For details see James M. Douie, *Punjab Settlement Manual*, 3rd edn (Lahore, 1915), pp. 245-6.
54. Trevaskis, p. 237.
55. *Report on Factors Affecting Irrigation in the Punjab From Tube Wells* (Lahore, 1938).
56. According to one estimate (*ibid.*, p. 25) net income per acre from tube-well irrigation was 45 per cent higher than the income from open wells.
57. For details see *ibid.*, p. 92.
58. Roberts and Singh, pp. 154-62.
59. For more details see Roberts and Singh, pp. 145-6.
60. For details see *DG Gurgaon*, 1910, p. 125-31.
61. Saini, pp. 203-4.
62. *DG Kangra*, 1924-5, p. 280.

CHAPTER 2

TRENDS IN AREA IRRIGATED FROM DIFFERENT SOURCES

In the last chapter we were concerned with a general description of the irrigation system—an historical outline of canal construction and colonization programme, water management, different methods of lifting water from wells and the technological improvement achieved in this field. The present chapter is an attempt to analyse the long and short-term trends in total irrigated area and area receiving regular water supply from individual sources. The long-term trends are intended to reveal the rate of progress achieved in the expansion of water supply as against population growth and extension of total cropped area. The short-term changes should enable us to make some idea about the timing of the acceleration or deceleration of area irrigated from a particular source as also the impact of such factors as rainfall and changes in agricultural price and water rates. Long-term trends have been estimated by fitting exponential equation of the type $\log Y = a + bt$. We propose to indicate the short-term changes with the help of decennial index numbers. An attempt is also made to indicate the sources of expansion of irrigated area and account for the disparity in the performance of the different sources of water supply over time.

2.1 Nature of the Statistical Data

The data base is drawn from two annual publications of the Punjab government—*Land Revenue Administration Reports* for the period 1887-8–1900-1 and *Season and Crop Reports* for the subsequent years. These publications reported irrigated acreage under the following heads: (a) public canals (b) wells (c) private canals (d) tanks and (e) other sources. But area irrigated from the last three sources accounted for only 5 per cent of the total irrigated area in 1901-2. Therefore, for the sake of convenience, area served by these three sources have been added up and described as area irrigated from miscellaneous sources. Area irrigated from Major and Minor (Public) canals was also reported by the provincial Irrigation Department, but there is a slight discrepancy[1] in the data available from the two sources for the period 1911-12–1946-7. This discrepancy was due to the fact that whereas the data returned by the revenue and agricultural officials "represent the canal-irrigated area actually sown", the Irrigation Department "takes no account of this aspect and includes area which receives canal water whether sown or not".[2] There is no discrepancy in the earlier years (i.e. prior to 1911-12) because statistics on area irrigated from public canals made available in the *Land Revenue Administration Reports* and *Season and Crop Reports* were those supplied by the Irrigation Department. In other words, the Agricultural Department did not itself collect statistics on area irrigated by Pubilc Canals during the earlier years. Thus, it is clear that this set of data is not uniform in quality during the entire period under review. But it has not been possible to attempt any adjustment on this count.

There are two other elements of non-comparability which are common in the case of statistics on all crop acreage and area irrigated from all the sources. Firstly, in November of 1901 the north-western parts of Greater Punjab were separated to constitute North-Western Frontier Province. This would not have created a problem if the whole of all the north-western districts were separated, for in that case data relating to these administrative units could have been deducted from the provincial (as it was constituted before 1901) total. But a survey of the relevant data shows that the partition was not carried out in this manner in the case of all the districts. Secondly, statistics on crop acreage and irrigated area were those on *harvested area* up to the year 1905-6 and *sown area* from 1906-7 onwards. To overcome these problems two sets of trends rates have been presented: one for the first 14 years (i.e. up to 1900-1) and the other for the period 1906-7–1946-7. In preparing the index numbers for the period 1887-1900 no adjustment has been made in the data on account of boundary changes. In other words, the acreage data of the last fourteen years of the nineteenth century relate to the Greater Punjab (i.e. British Punjab and North-Western Frontier Province). So far as the second element of non-comparability is concerned, we have tried to overcome it by working out the average of the fourth decade from 1906-7. On the other hand, the fifth decennial average is the average of 11 years (1926-7– 1936-7). This has been done is order to highlight the impact of the Great Depression.

Before we proceed farther, a brief reference needs to be made to the larger question of the reliability of the statistics on crop acreage and yield per acre—a subject which exercized the mind of the members of the colonial bureaucracy in the past and has continued to be discussed by the present-day scholars.[3] This discussion is relevant in the present context because these scholars are not concerned with the quality of only the data on yield rates and cropped area which did not receive artificial water supply, but also those on yield and sown area irrigated from different sources, private and public. Broadly speaking, three views have emerged from this discussion. Firstly, it is generally believed that the acreage statistics reported by the *Patwaris* are reliable. For example, Panse who was appointed by the British government to formulate a plan for improving the crop statistics opined that except in the permanently settled areas "acreages under crops are known with a high degree of accuracy".[4] Secondly, yield rates of the individual crops as worked out by the Agricultural Department were underestimates. As pointed out in an official publication, "while we have a very accurate knowledge of the acreage, the estimates as to yield must be accepted with considerable hesitation. The principal exporting firms, for example, get fairly good results by assuming that wheat outturn is 33 per cent greater than official estimate."[5] Similarly from the returns submitted by the factories under the provisions of the Ginning and Pressing Factories Act (1925) it was found that Standard Yield of cotton during the quinquennium 1927-8–1931-2 was underestimated by 28.5 per cent.[6] The findings of other enquiries indicated the same pattern, and accordingly in

1938-9 the Standard Yields of *deshi* and American varieties of cotton were revised from 123 and 130 lbs to 190 and 192 lbs, respectively.[7] Thirdly, a more complicated question has been raised by Clive Dewey. According to him the yield data worked out by the official agencies were not only underestimates, but the range of error increased over time.[8] A close look at the yield figures of crops like wheet, sugarcane and cotton published by the Department of Agriculture and progress achieved in the spread of new varieties of seeds of these crops does actually suggest that Dewey's view has to be taken seriously.[9]

From the foregoing discussion it is clear that while trend rates in cropped area irrigated from different sources of water supply can be estimated on the basis of the available statistics, some adjustment is necessary in the case of yield figures. This is because the degree of underestimation in these statistics increased over time. But obviously it will be extremely difficult to make some idea about the increase in the margin of error over time. Therefore, the trend rates in irrigated acreage and yield rates have been estimated on the basis of the available data.

Finally, it needs to be emphasized that while some considerations suggest that the estimated trend rates in crop acreage served by canals will exaggerate the actual irrigation capacity created in this sector over time, others will have the opposite effect. With regard to the former, it is quite plausible that the existing canals functioned more efficiently in the sense that water supply at the canal head was augmented and/or wastage in the distributaries reduced. It is also likely that the irrigators used the same quantum of water for cultivating a larger area. In such a situation area irrigated would expand faster than actual irrigation capacity. On the other hand, greater emphasis on the production of crops which required a larger number of waterings would underestimate the progress achieved. We may, however, proceed on the assumption that such factors will not constitute a serious obstacle in the way of taking the estimated trend rates irrigated by canals as an index of the expansion of the irrigation capacity in the public sector.

2.2 Rates of Change During 1887-8–1900-1

Estimated annual rates of change as presented in Table 2.1 indicate that area watered by wells slightly declined during the period 1887-8–1900-1, but canal irrigation expanded rapidly, at the rate of 6.5 per cent per year. In absolute terms this increase represents an addition of 3.2 million acres to the area (2 million acres) irrigated from this source in 1887-8. Given such a phenomenal achievement in this sector it is not surprising that the Indian Irrigation Commission (1903) commented: "Although little more than half a century has elapsed since the Punjab came under the British rule, it is here that the greatest progress has been made in irrigation Works. . . ."[10] The expansion rate could have been still higher but for a certain shrinkage in canal-irrigated acreage over two consecutive years (1892-3 and 1893-4) due to unusually favourable rainfall. Anyway the explanation of this impressive performance is not far to seek. For, as mentioned earlier, Punjab received lion's share of the fund borrowed for investment in canal-building activities and the last two decades of the nineteenth

TABLE 2.1: ESTIMATED TREND RATES IN IRRIGATED ACREAGE

Sources of Water Supply	*1887-8–1900-1*	*1906-7–1946-7*
Canals	6.50	1.80
Wells	-0.35	0.84
Miscellaneous	-0.45	-0.23
Three source together (i.e. total)	2.41	1.42
Unirrigated area	3.06	-0.19
Unirrigated + irrigated area	-0.74	0.65

Source: Data presented in Appendix.

century saw the opening of the Sirhind, Lower Sohag and Para, Sidhnai and Chenab Canals and the Sirsa Branch of the Western Jamuna Canal.

The rapid expansion in irrigation from public canals was accompanied by some decline (0.35) in area served by wells and miscellaneous sources of water-supply (embankments, tanks, private canals, etc.). The fact that this marginal decline in irrigation from private sources has such marked impact on the trend rates in total irrigated area (2.41 per cent per year) is not difficult to explain. For area irrigated from wells and miscellaneous sources accounted for 64 per cent of the total irrigated area in the province in 1887-96. However, though the decline in irrigation from private sources lowered the expansion rate in total irrigation area during the first fourteen years of the period under review, the latter still expanded much faster than population growth in the province (0.85 per cent per year).[11]

The contraction of unirrigated area during these years needs emphasizing as, otherwise, this might give the impression that the process was more or less the same as in the United Province where canal irrigation expanded primarily in areas with settled agriculture. In other words, it might appear that canal irrigation expanded in the Punjab not so much because virgin lands were brought under the plough, but primarily as a result of water supply to areas already under cultivation. But it has to be recalled here that, unlike in the United Province, the larger part of the tracts commanded by the newly built canals in the Punjab were (with fertile land but inadequate and uncertain rainfall) such that these hardly supported any settled population. This pattern necessitated a massive colonization programme and the consequent shift in population density in different parts of the province (Table 2.2). It is true that part of the unirrigated crop acreage was now commanded by the expanding network of canals, but this does not account for the larger part of the decline. This is obvious because in absolute terms the decline under this head (unirrigated area) far exceeded the expansion in canal irrigated crop acreage. Actually the larger part of the contraction was due to the fact that rainfall was unusually unfavourable in three years—1896-7, 1897-8 and 1900-1. Incidentally unfavourable weather conditions did not only hinder sowing, but increased the rate of crop failure. Indeed in 1900-1 harvested area declined to such an extent (harvested area was only one-sixth of what it was in the first quinquennium) that even some expansion in irrigated area could not prevent the outbreak of a famine in the province.

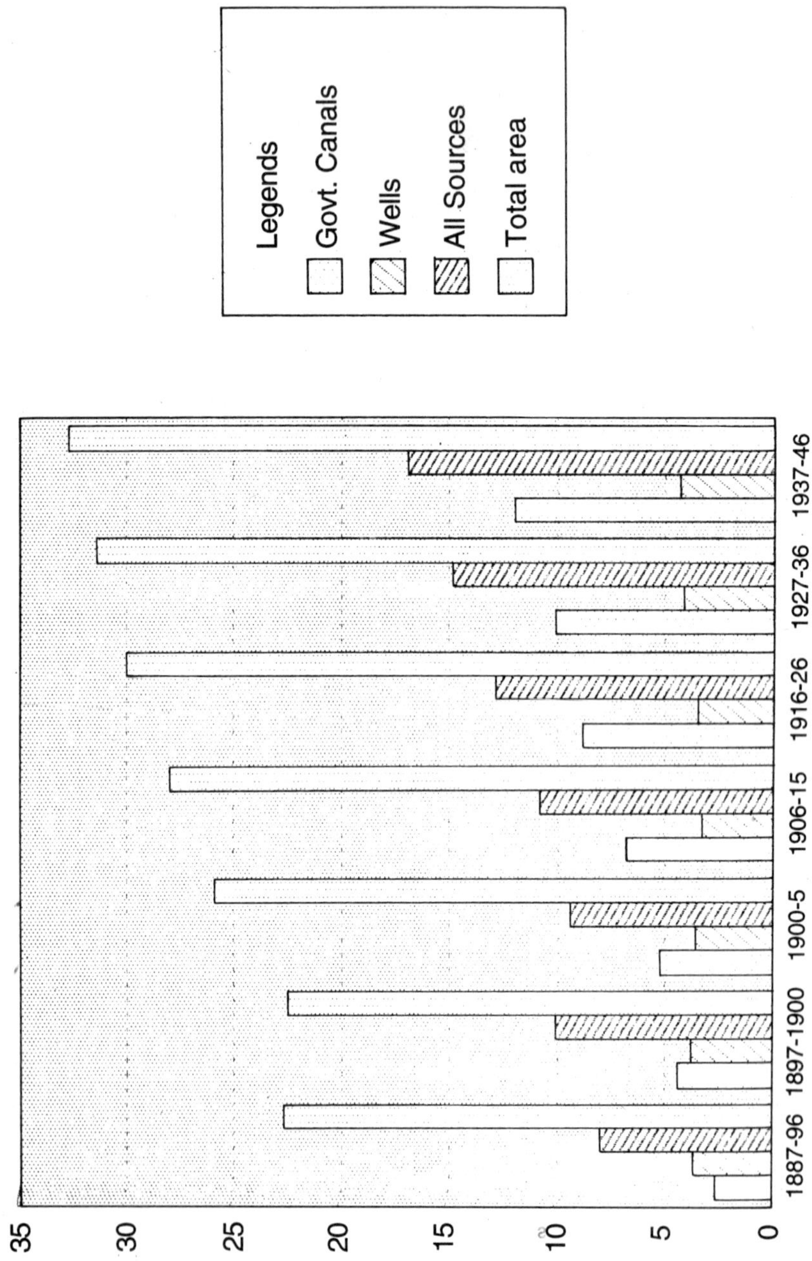

GRAPH 1: AREA IRRIGATED FROM DIFFERENT SOURCES (MILLION ACRES)

TABLE 2.2: PUNJAB: VARIATION IN POPULATION, 1891-1941

	Number of Persons ('000)						*Per Cent Variation*
	1891	*1901*	*1911*	*1921*	*1931*	*1941*	*1891-1941*
Ambala	865	817	691	681	743	848	1.96
Amritsar	993	1,024	881	930	1,117	1,414	42.38
Attock	448	464	519	512	584	676	50.72
Dera Ghazi Khan	428	472	500	469	491	581	35.71
Ferozepore	886	958	960	1,099	1,157	1,423	60.60
Gujranwalla	661	740	606	624	736	912	37.96
Gujrat	806	792	788	824	922	1,105	37.03
Gurdaspur	944	940	837	852	971	1,154	22.20
Gurgaon	760	843	730	682	740	851	12.00
Hissar	776	782	805	817	899	1,007	29.75
Hoshiarpur	1,011	990	919	927	1,932	1,170	15.71
Jhang	402	426	525	571	665	882	104.21
Jhelum	514	501	512	477	541	630	22.48
Jullunder	908	918	802	823	944	1,127	24.20
Kangra	763	768	770	766	801	899	17.87
Karnal	863	885	801	829	853	995	15.31
Lahore	899	1,004	1,000	1,130	1,379	1,695	88.63
Ludhiana	649	673	517	568	672	819	26.19
Lyallpur	60	586	836	968	1,167	1,396	2,215.37
Mianwali	287	302	341	358	412	506	76.40
Montgomery	417	430	482	686	1,000	1,329	219.07
Multan	621	700	801	879	1,160	1,484	138.96
Muzaffargarh	494	528	569	568	591	713	44.33
Rawalpindi	534	559	548	569	634	785	47.12
Rohtak	780	834	715	772	806	956	22.63
Shahpur	478	488	645	720	821	999	108.85
Sheikhupura	401	523	540	634	696	853	112.46
Sialkot	962	933	872	878	980	1,190	23.80
Simla	36	39	38	45	37	39	6.85
Total Punjab (British Territory)	18,653	19,943	19,579	20,685	23,419	28,419	52.36

Source: *Census of India*, 1914, Vol. VI (Punjab), pp. 8-11.

However, as implied earlier, it is clear that decline of area watered from private sources—wells, tanks, embankments and private canals has to be entirely attributed to the expansion of canal irrigation. The magnitude of replacement was not as marked as in the United Province. Moreover, the substitution did not continue throughout the period under review. But canal irrigation did partly replace irrigation from wells (and other private sources of water supply) and, as indicated by the evidence available from various government reports, this caused concern among the officials. In some areas the trend was so marked that they suggested specific steps to arrest it. For example, Major J.W. Ottley, the Superintendent of Doab Circle, suggested that canal irrigation within definite well-*chaks* should be absolutely prohibited or well land should not be opened to canal irrigation without the payment of double charges.[12] How can we explain this substitution? Was this switch voluntary or was it imposed on them? These questions deserve attention particularly because yield per acre of well-lands was considerably higher than that of lands watered by canals. Indeed according to one estimate an acre of well-irrigated land produced at least one-third more than a similar land served by canals.[13]

According to one view cultivation with the help of canal water was not so laborious as *chahi* cultivation and therefore it was more congenial for the "indolent" peasants.[14] It is quite probable that there were peasants who were ready to spend efforts only up to the point where a certain "target" income was attained and, accordingly, they abandoned well-irrigation when canal water became available. But it would be unrealistic to generalize on the basis of such evidence. The Indian Irrigation Commission opined that the low state of the rivers or the absence of heavy spills reduced the spring level in the vicinity of the wells. Therefore, as it became necessary to apply a larger number of waterings to the fields, the cultivators turned to the canals which could supply the required quantum of water.[15] This would mean that as far as the proprietary villages (village with settled population) are concerned canal irrigation was imposed on the peasantry. Diversion of river water through the newly-constructed canals and their distributaries must have created such a situation in some parts of the province. But once again it would be difficult to believe that this was the general pattern. Incidentally, we have evidence to suggest that in certain tracts the rise in the sub-soil water caused by the operation of the canals made the working of the wells easier.[16] According to yet another view, well-irrigation declined because canal water was much cheaper.[17] Canal irrigation did really have this advantage but this factor does not, by itself, explain much. For example, it does not explain why the cultivators turned to canals even though this meant lower yield per acre.

As a matter of fact it seems that the explanation for the preference of the cultivators for canal-irrigation lies in the fact that this technology had certain advantages over well-irrigation. As mentioned in the previous chapter, *Charsa* was the most important method of lifting water from the wells till the early years of the twentieth century. But compared to irrigation from canals this method of irrigation was more labour-intensive, given that two or three able-bodied men and six to eight bullock were needed to operate a *charsa*. Secondly, the irrigating capacity of such wells was limited even in the tracts most favourable for well-irrigation. This was because the quantum of water which could be lifted per operation was limited and the process of lifting water was time-consuming. It may also be noted in this connection that these wells could not always be relied upon for adequate water supply. Even if the supply could be sustained, bullocks could not work for more than a certain number of days in the month. In all these respects canal-irrigation represented a marked innovation. Firstly, as the canal water flowed on to the fields by the force of gravity the requirement of labour, both human and bullock, was much lower and the process of watering was easier and less time-consuming. Secondly, if a certain tract was not unfavourably situated at the farthest end of a distributary or if there was no drought, canals could be relied upon for adequate supply of water. Thirdly, the cultivators had to pay for canal water according to crop acreage, i.e. area irrigated, not by the volume of water used. But the water rate was moderate and consequently the cultivators was left with a considerable part of the output accruing from the use of canal water.

To summarize, then, canal-irrigation meant less yield per acre but offered scope for more extensive and less expensive cultivation. On the other hand, though well-irrigation ensured higher yield rate, it also entailed higher cost of production and more intensive cultivation. But given that there was scope for the extension of irrigation even in the proprietary villages (i.e. older districts) in the late nineteenth and early twentieth centuries these advantages of wells were outweighed by those of canals. As canal water was much cheaper and adequate in supply, the cultivators had a tendency to extend irrigation not only to the fields hitherto unirrigated and even uncultivated but, wherever possible, to switch from wells to canals for water supply. Such a substitution meant lower yield per acre, but this loss was more than neutralized by the addition to total production now made possible by the increase in the proportion of irrigated area in the cultivators' holdings. In the circumstances some wells went out of use (and some were used to supplement canal water), thus leading to an absolute decline in well-irrigation.[18] Needless to mention, replacement of wells as a source of water-supply meant the release of the factors (human labour and bullocks)[19] which, in association with canal water, could expand the irrigated acreage. Overall, substitution of well-irrigation by canal water, to the extent that it did happen in some areas, was due to the latter's relative advantages.[20]

2.3 Trend Rates During 1906-7–1946-7

Trend rates (Table 2.1) during the four decades from 1906-7 have several important features. Firstly, irrigated area continued to expand, but on a much reduced scale. Indeed the annual expansion rate (1.80 per cent) was a little over one-third of the rate (6.50 per cent) achieved during the first two decades. In absolute terms this expansion rate meant an addition of another 6,000,000 acres to the irrigated area. The explanation for this deceleration lies partly in reduced investment by the government and partly in the problem of waterlogging. With regard to the first, it may be recalled that after the opening the Upper Jhelum Canal in 1915-16, the only new canals to come into operation in the province were the Pakpattan, Mailsi, Dipalpur and Eastern Canals. Incidentally it should also be mentioned that irrigation became a "reserved" provincial subject under the Indian Council Act of 1919. The impact of the problem of waterlogging began to be felt from the turn of the century and by 1946-7 the problem became so acute that one million acres of canal-irrigated area in the thirteen districts of West Punjab went out of cultivation. Secondly, the downward trend in well-irrigation was now reversed. The expansion rate was much lower in this field, but there was now definitely an improvement—at the provincial level well-lands increased by about one million acres between 1906-7 and 1946-7. Thirdly, contraction of the area irrigated from miscellaneous sources (tanks, embankments and private canals) continued during the second phase. Fourthly, area irrigated from all three sources taken together expanded faster than total cropped area and also exceeded population growth. The effects of the diverse performance

TABLE 2.3: STATISTICS ON AREA IRRIGATED FROM DIFFERENT SOURCES

Years	*Govt. Canals*			*Wells*			*Misc. Sources*			*Total*		
	a	*b*	*c*	*a*	*b*	*c*	*a*	*b*	*c*	*a*	*c*	*d*
1887-96	2.6	36	12	3.7	50	16	1.7	14	5	7.4	33	-
1897-00	4.4	50	22	3.8	41	19	1.9	12	5	9.3	47	-
1901-5	5.2	55	20	3.6	39	14	0.6	7	2	9.3	37	-
1906-15	608	63	24	3.3	31	12	0.7	6	2	10.8	39	105
1916-26	8.8	69	30	3.5	27	12	0.5	4	2	12.8	44	125
1927-36	10.1	68	32	4.1	28	13	0.6	4	2	14.8	47	143
1937-46	12.0	70	37	4.3	26	13	0.6	4	2	17.0	53	165

Source: Computed from data in the Appendix.
Notes: (a) Area in million acres (b) as a percentage of irrigated area (c) as a percentage of total sown area and (d) Index number computed with 1906-7–1910-11 as the base. 1887 stands for 1887-8 and so on.

of the different irrigation technologies are shown by data presented in Table 2.3. Wells lost their share of the total irrigated area by half and the miscellaneous sources by about one-fourth. On the other hand, the government canals doubled their share during the six decades.

With regard to short-term trends, canal-irrigation expanded more or less in all the four decades (Table 2.3) of the twentieth century. The expansion during the decade 1906-7–1915-16 was primarily due to the opening of these canals—Upper Chenab, Lower Bari and Upper Jhelum. Another impetus was provided by the upward trend in agricultural prices which set in at the turn of century. Water supply was abnormally low in 1907-8 and, to some extent, in the following year; but as far as the expanstion rate is concerned these losses were more than neutralized by the additional area served by these canals. Growth rate during the next decade (1916-17–1925-6) was the most impressive and this is explained by two favourable factors. Firstly, this decade saw the opening the several new canals—Pakpattan, Mailsi, Dipalpur and Eastern. Secondly, the upward shift in agricultural prices gathered momentum during these years. For example, the price of cotton registered an increase of about 100 per cent from 1915-16 to 1925-6. During the same period the price of wheat increased by 50 per cent. Expansion of canal-irrigated area continued during the Great Depression (1926-7–1935-6), thus belying the prediction of the Abiana Committee that "unless rates are reduced the Irrigation Department is sure to loose a part of its customer. The agriculturist will begin to replace high-rated crops by low rated crops and, generally, to content himself with a little irrigated area as possible".[21] It may be mentioned here that agricultural prices had begun to decline from 1926-7 and this almost coincided with the imposition of an enhanced water rate in the province. But there was no downward trend in canal-irrigated area. Then as the Great Depression set in all over the world and agricultural prices further declined the government granted special rebate on water rate from 1930-1 and finally, from 1933-4, reduced the rates by about one-fifth.

But the area under canal-irrigation further expanded even though the reduction in water-rate did not at all match the fall in agricultural prices. As a matter of fact acreage served by government canals during 1927-8–1936-7 exceed the area irrigated over the preceeding 11 years by 15 per cent. We propose to attempt a detailed analysis of the water rates at a later stage, but from the pattern that emerges here it is clear that water-rate was not the most important determinant of the cultivators' decision about the use of canal-water. Expansion rate accelerated during the next decade (1937-8–1946-7). The officials of Agricultural Department attributed this developement to the success of the Grow More Food Campaign launched by the government during the early 1940s and the sharp rise in agricultural prices. It is possible that the Grow More Food Campaign did play a certain role but the price factor seems to have provided the greater stimulus to the marked expansion of crop acreage drawing water from public canals.

The decennial trends in well areas have to be interpreted with reference to nature of the data mentioned earlier. The statistics relating to Punjab proper (i.e. without the districts which constituted the North-Western Frontier Province) are not strictly comparable in the sense that the data reported during the first years (1901-2–1905-6) are those on harvested area and those on the subsequent years on sown area. However, given that harvest hardly failed in lands served by wells this element of non-comparability will not pose a serious problem. It may then be suggested that the replacement of well-irrigation which started at the outset of the period under review continued till 1915-16. The trend was reversed during the next decade, but expansion rate was very slow. Thus, the impact of rising agricultural prices was confined to canal-irrigation only. Well lands expanded much faster during the Depression years and consequently the area now served by this irrigation technology was 20 per cent higher compared to the level of 1901-2–1905-6. Thus, the pattern in well-irrigation was now the same as in canal-irrigation—well areas continued to expand even though agricultural prices ruled very low. Well lands further expanded during the last decade 1937-47.

At first sight it might, then, appear that the argument developed earlier to explain the replacement of well-irrigation during the first phase does not really hold good. But a closer look makes it clear that this is not really the case. As a matter of fact in a way the pattern in well-irrigation during the last two decades reinforces our contention that the cultivators' preference for a particular irrigation technology was, by and large, dictated by its relative advantages. As shown earlier, irrigation with the help of *Charsa* had certain basic disadvantages. But these were being overcome during the last two decades as a result of certain improvements in the method of lifting water from the wells. Firstly, Persian Wheel (*Rahat*) which was a more efficient water lift, in the sense that it was less labour-intensive and capable of supplying a larger volume of water, was replacing *Charsa*. Secondly, some progress was being made in augmenting water supply by boring wells and installing suitable engines and pumps. It may here be mentioned that according to the Live-Stock Census Report, 504 oil engines with pumps for irrigation purposes and 268 electric pumps for tube-wells were in operation in the province in 1940.[22] Finally, it is quite plausible that during the Depression years many cultivators made greater use of the wells which were already in service or brought back into service some of wells which had remained out of use. This is because given the fall in agricultural prices they were under pressure to reduce the paid-out cost of cultivation. In this connection it may also be added that, as the Depression continued, cash wage rate of hired labour also declined. Moreover, it is not unlikely that a section of the cultivators fell back on the family labour which had earlier ceased to take part in agricultural operations.

2.4 Regional Distribution of Irrigated Area and Sources of Expansion

There were considerable variations in the spread of canal and well-irrigation (Table 2.4) in different parts of the province. As can be expected from the nature of the canal construction programme of the government, canal-irrigation was most prominent in the colony districts—Lahore, Multan, Gujranwalla, Jhang, Layallpur, Sheikhpura, Shahpur and Montgomery. All these districts later came to be included in the West (or Pakistani part of the) Punjab. Muzaffargarh, another district where canal water served more than 50 per cent of the sown area, was also included in West Punjab. Together these districts accounted for 75 per cent of the total canal irrigated (and 35 per cent of the total sown) area of the province. Incidentally from the statistics on the proportion of culturable waste lands it is clear that in most of these districts there was still considerable scope for the expansion of cultivation. Among the districts which were later included in the Indian part of the Punjab it was only in Ferozepore and Amritsar where the proportion of canal irrigated area exceeded the provincial average (35 per cent). But together these two districts accounted for only one-tenth of the total area of the province served by government canals.

Distribution of well-lands has three important features. Firstly, the districtwise distribution was less unequal—it was only in one district (Jullunder) that more than 50 per cent of the total sown area was irrigated by wells. Secondly, in most of the districts in which well-irrigation played a prominent role the scope for the further expansion of cultivation was very limited. Thirdly, a glance at the districtwise data suggests that the Indian part of the Punjab included the larger part of the well-lands, but the disparity was not as marked as in the case of canal-irrigation.

The most important source of expansion of irrigated area was land designated in the official publications as "culturable waste" (Table 2.5). This is obvious because the government irrigation works were primarily constructed in tracts with little settled pupulation but plently of cultivable land. The second most important source was the area which was hitherto cultivated but did not receive regular water supply. The third source was the lands described in the official publications as "current fallow". It may be recalled here that area actually irrigated by the canals always exceeded the area "proposed (by the Irrigation Department) to be irrigated". Fallowing may also have declined in the predominantly well irrigated areas, but it is likely that this trend was more marked in the canal lands. Incidentally, according to one view reduction in fallowing and replacement of the traditional leguminous rotation and the expansion of the area under fodder (such as *alfalfa* and Egyptian clover which are more powerful

TABLE 2.4: DISTRICTWISE STATISTICS ON IRRIGATED AREA, ETC.
(AVERAGE OF 1941-2 AND 1942-3)

Districts	*P.C. of Area Canals*	*Irrigated by Wells*	*P.C. of Area Culturable*	*Current Fallow*
Hissar	12	00	11	15
Rohtak	20	06	13	04
Gurgaon	07	07	10	05
Karnal	20	14	31	09
Ambala	00	04	10	04
Kangra	00	00	00	01
Hoshiarpur	00	06	15	02
Jullunder	00	53	09	03
Ludhiana	12	37	08	04
Ferozepore	43	09	09	08
Lahore	66	17	15	05
Amritsar	42	38	12	03
Gurdaspur	08	17	08	02
Sialkot	00	40	08	03
Gujranwalla	34	38	23	04
Sheikhpura	64	14	25	05
Gujrat	29	12	14	05
Shahpur	49	07	20	05
Jhelum	00	00	07	03
Rawalpindi	00	00	08	02
Attock	00	00	08	03
Mianwali	00	03	47	03
Montgomery	84	10	22	09
Lyallpur	91	03	21	06
Jhang	59	25	43	06
Multan	87	01	29	10
Muzaffargarh	57	17	64	02
D.G.Khan	42	04	45	05
Total	35	13	22	05

Source: Statistics on irrigated area are drawn from *Season and Crop Reports* and those on culturable waste and current fallow from *Agricultural Statistics of India*, Vol. 1.

legumes) meant that the traditional soil fertility maintenance mechanism was undergoing a change in the Punjab canal colonies.[23]

2.5 Problem of Waterlogging

A fuller discussion on the problem of waterlogging is outside the scope of this study. However, a brief reference will be in order given the considerable adverse impact it had on the agrarian economy as well as the human environment in the province. The first point that deserves emphasis in this connection is that engineers who designed

TABLE 2.5: SOURCES OF EXPANSION OF IRRIGATED AREA ('000 ACRES)

As Designated As	*1906-7–1909-10*	*1941-2–1943-4*	*Difference*
Culturable Waste	17,029	13,422	3,607
Unirrigated	17,788	16,066	1,722
Current Fallow	3,581	2,622	959
Failed (a)	1,066	697	369
Total Irrigated	10,550	17,207	6,643

Source: Computed from data in the Appendix
Note: (a) Irrigated area in which harvest failed

the earlier canals were aware of the need for taking steps to avoid the problem of waterlogging. For, as mentioned earlier in the first chapter, they laid down a formula according to which the proportion of area to be irrigated by a particular canal was to be determined keeping in view the spring water level below the surface. It was widely known that the area actually irrigated by a canal was exceeding the limit stipulated by the Irrigation Department.[24] However, when waterlogging became a matter of public concern at the turn of the present century the opinion was formed that percolation from canals, distributaries and water-courses was primarily responsible for this evil.[25] Accordingly, the first steps taken by the government were directed to lining the canals with water proof material. This opinion was formed not so much on any scientific investigation into the problem of waterlogging *per se,* but on the basis of the widely held view that percolation rate from the Punjab canals was very high.[26]

However, as waterlogging continued to affect new areas another opinion began to emerge—the problem was caused by the tendency of the cultivators to over-irrigate their lands. It was argued that "the absurdly low water rates and the complete absence of trouble (on the part of the irrigators) in getting water led to persistence and considerable wastage of water in the field and this led to waterlogging problem".[27] Many other commentators also took the view that the cultivators were applying more canal water per unit of land and waterings were more frequent during the life of crops than was necessary for successful cultivation, and all of them attributed this tendency to the fact that canal water was not priced according to volume supplied but by area irrigated.[28] The only check on wasteful consumption was the time allowed for each irrigator, but this proved insufficient to ensure economic utilization of water. Some inquiries did actually show that area irrigated by a given volume of water lifted by Persian Wheel was much in excess of that irrigated by flow from an outlet in canal.[29] The only dissenting voice came from W.P. Sangster, Chief Engineer, Irrigation Works, Punjab. He told the Royal Commission on Agriculture that the quantity of water allowed per outlet was so "limited that really they (irrigators) are forced themselves to exercise as much economy as possible".[30] However, when the subject of waterlogging came to be reviewed at a conference in Lahore in April 1917, it was agreed that the rise in the sub-soil water table was more

TABLE 2.6 : WATERLOGGED AREA ACCORDING TO ORDINARY GURDAWARI ('000 ACRES)

Years	*Acres*	*Years*	*Acres*
1927-8	284.9	1937-8	483.4
1928-9	272.4	1938-9	513.4
1929-30	145.2	1939-40	530.6
1930-1	327.6	1940-1	552.7
1931-2	224.3	1941-2	628.6
1932-3	347.5	1942-3	712.9
1933-4	348.3	1943-4	1,414.6
1934-5	363.6	1944-5	1,643.9
1935-6	377.3	1945-6	1,854.4
1336-7	438.1	1946-7	1,945.2

Source: Punjab Board of Economic Inquiry, *Agricultural Statistics of the Punjab, Pakistan* (Lahore, 1950).

due to the percolation from channels than to the water spread over the fields and it was more important to avoid obstruction to natural drainage than to restrict irrigation.[31]

Until 1920 waterlogging question was dealt with by the Irrigation Department, but in that year the provincial government established the Drainage Board (later called Rural Sanitary Board) and entrusted it with the supervision and coordination of efforts for prevention and cure of waterlogging problem. In 1925 a Waterlogging Enquiry Committee was constituted to investigate into the relative role of such factors as (a) irrigation of the field (b) seepage from canals (c) rainfall and flood (d) blocking of surface channels to the rise of subsoil water level. Meanwhile, the government continued taking such steps as digging of scapage drains, clearance of natural drainages, closure of certain canals/distributaries at certain times of the year etcetera.[32] Between 1920-1 and 1933-4 a total of little over 10 million rupees was spent by the provincial government to tackle the waterlogging problem.[33] Incidentally, as elaborated later, the Punjab government were also examining several proposals to introduce volumetric assessment of canal water.

But these measures proved inadequate and the problem of waterlogging became more acute over time—waterlogged area standing at about two million acres in 1946-7 in the thirteen districts of West (Pakistan) Punjab. The problem was even worse according to a special *Gurdawari* which showed that 2.4 million acres (or about one-fourth of the total canal irrigated area) were affected in these districts in 1946-7 and out of these one million acres had gone out of cultivation.[34] Secondly, waterlogging had considerable adverse impact on the yield rates of crops, especially of cotton crop.[35] Thirdly, waterlogging led to a deterioration in the human environment in the province, especially evidenced by the high incidence of malaria epidemic. This is obvious from the "final conclusions arrived at" by the provincial Director of Public Health,[36]

> . . . open field irrigation has not been responsible for any appreciable general increase in endemic malaria; . . . (but) as a partial exception to general rule it is certain that wherever canal irrigation gives rise to waterlogging a vicious circle sets up in which endemic malaria leads to bad health, bad health to economic stress, and economic stress to further privation and more sickness, and finally, as the result of a high death-rate and low birth-rate, emigration and depopulation of the affected area.

It may be added here that during 1925-30 fever death-rate, owing chiefly to an epidemic of malaria, accounted for two-thirds of the death-rate in the province.[37] Secondly, Punjab province had the highest death-rate in the subcontinent.[38]

Thus, to summarize, canal-irrigation expanded throughout all the six decade under review, but the expansion rate was much faster during the first two decades. On the other hand, well lands contracted these years, thus indicating that well-irrigation was being partly substituted by canal-irrigation. By and large this substitution was dictated by the fact that in the given situation in the Punjab province canal-irrigation was found more suitable. The intrusion of canal-irrigation into well areas soon came to an end and there was a certain expansion of area served by wells during the last two decades. Irrigation from other sources (tanks, private canals, embankments) remained stable during the first two decades, but declined in the subsequent years. Overall, area served by different sources of water supply increased by more than seven million acres during 1905-47. The expanding network of canals and distributaries brought in its trails the problem of waterlogging with its adverse impact on agricultural production and human environment, but the contribution of the canal construction programme of the government to the development of the vast irrigation system that Punjab came to possess is not in doubt.

NOTES

1. The discrepancies ranged from 300 and 1,817,000 acres.
2. *Agricultural Statistics of the Punjab* 1901-2–1935-6 (Board of Economic Inquiry Publication No. 52, Lahore, 1937), p. 11.
3. Clive J. Dewey, "The Agricultural Statistics of the Punjab, 1867-1947", *Bulletin of Quantitative and Computter Methods in South Asian Studies,* No. 2 (March 1974); "*Patwari* and *Chaukidar*: Subordinate Officials and Reliability of India's Agricultural Statistics" in Clive Dewey and A.G. Hopkins (eds.), *Imperial Impact: Studies in the Economic History of Africa and India* (London, 1978); S.C. Misra, "On the Reliability of Pre-independence Agriculture Statistics of Bombay and Punjab", *Indian Economic and Social History Review*, XX (2) and C.E. Pray, "Accuracy of Agricultural Statistics and the Sources of Growth in the Punjab", *Indian Economic and Social History Review*, XXI (2).
4. V.G. Panse, *Report on the Scheme for the Improvement of Agricultural Statistics* (Simla, 1946), p. 14.
5. *Season and Crop Reports* (1922-3).
6. *Report on the Marketing of Wheat in the Punjab* (Lahore, 1939), p. 27.
7. *Annual Report of the Indian Central Cotton Committee*, 1945, p. 72.
8. Clive Dewey, *op. cit.*
9. M.M. Islam, "Trends in Crop Production in Undivided Punjab: A Reassessment" in C.J. Dewey (ed.), *Arrested Development in India: The Historical Dimension* (New Delhi, 1988).
10. *Report of the Indian Irrigation Commission,* 1901-3, Part 2, Provincial (London, 1903), p. 1. Given that canals in this province proved so profitable and there was enormous scope for the further

expansion of canal-irrigation the Commission recommended the investment of Rs 11,000,000 in canal construction during the next 20 years. For further details see Part 1, p. 41.

11. Population growth rates in the Punjab were as follows: 1871-81 + 0.6 per year; 1881-91 + 1.01 per cent; 1891-1901 + 0.69 per cent; 1901-11 minus 0.18 per cent; 1911-21 + 0.57 per cent; 1921-31 + 1.40 per cent and 1931-41 + 2.05 per cent.
12. Land Rev. & Ag. (Rev.), March 1891, A, No. 27, p. 11. His contention was that canal water which was unnecessarily used in the well-tracts could, with greater advantage to community, be passed on to the dry areas where well-irrigation was impossible. Needless to mention, this would have also increased government revenue. Colonel Grey, Superintentent of Irrigation, Bhawalpur State, suggested that canal water should by given so sparingly as to drive the people to use the wells they had and to sink others. For details see *Report of the Indian Irrigation Commission*, 1901-3, Part III (Evidence), p. 6.
13. H.G. Trevaskis, *The Punjab of To-day*, Vol. 1, (Lahore, 1931), p. 239. It will be naturally difficult to say whether yield rates of well-lands were so much higher. But it can be safely assumed that the basic contention is valid. For the unit cost of water obtained from wells was higher and therefore the supply was carefully applied to the limited area it would cover and such land was much more meticulously cultivated.
14. *Assessment Report* (henceforth *AR*) *Mamdot Jagir, Ferozepore District*, 1913, p. 13. The same view was taken in *DG Jhang District*, 1929, p. 116. It was argued that the ease-loving tenants who flocked to the areas under canal-irrigation prefer to part with a greater share of the produce than put in hard labour on wells. This tendency was so marked that in many cases the tenants were found to refuse to work wells unless *Nahri* or *Sailab* lands were given to them in addition to *Chahi* lands.
15. *Indian Irrigation Commission*, 1901-3, Part 2, Provincial, p. 31. The result must have been the same in areas where canals rendered the sub-soil unstable, thus causing wells to fall in.
16. *Final Settlement Report of the Canal-Irrigated Tracts of the Gujrat District* (Lahore, 1930), p. 5.
17. *Ibid.*, p. 15; *DG Gujranwalla*, 1935, p. 120.
18. We are not trying to explain why well-irrigation did not further expand, but it was being replaced by canal-irrigation. Therefore, no reference is being made to the fact that, unlike irrigation from canals by flow, well-irrigation involved certain overhead charges (capital cost of sinking wells, installation of lift, expenditure incurred in connections with repairs and maintenance). For details of such costs see Sir William Roberts and S.B.S. Kartar Singh, *A Text-Book of Punjab Agriculture* (Lahore, 1947), pp. 154-62.
19. The numer of plough-cattle (bull, bullocks and male buffaloes) decreased from 19 per 100 acres sown in 1904 to 14 per 100 acres in 1935. For details see, Punjab Board of Economic Inquiry, *Agricultural Statistics of the Punjab, Pakistan* (Lahore, 1950), p. 27.
20. That decline in well areas was primarily due to economic consideration was admitted by some observers of the agrarian scene. For example, see *DG Gujranwalla*, 1935, p. 121.
21. *Report of the Abiana Committee* (Lahore, 1934), p. 26.
22. *Punjab Live-Stock Census*, 1940 (Lahore, 1940).
23. Hirashima Shigemochi, *The Structure of Disparity in Underdeveloped Agriculture: A Case Study of Pakistan Punjab* (Tokyo, 1974), pp. 33-4.
24. During the last decade area proposed to be irrigated exceeded the area actually irrigated by about one-third. For details see, *Administration Report of the Punjab Public Works Department* (Irrigation Branch).
25. W.P. Thompson, *Punjab Irrigation* (Lahore, 1925), p. 107. The waterlogging problem is also discussed in Punjab Irrigation Branch Papers issued at different times; H.K. Trevaskis, *The Punjab of To-day: An Economic Survey of Punjab in Recent Years, 1890-1925* (Lahore, 1931), pp. 299-305; M.T. Gibbling, *Concise Report on the Punjab Irrigation Works* (Bombay, 1932); *Report on Waterlogging in Punjab* (Lahore, 1924); Waterlogging Board Punjab: *Proceedings of Meetings*, Vols. 1-5.
26. In the case of the Upper Bari Doab Canal it was reported (*Report on the Rayal Commission on Agriculture in India*, Vol. VIII: *Evidence Taken in the Punjab*, pp. 4-5) that out of every 100 cubic feet entering the head of the canals 20 feet were lost in the canals and branches, 6 feet in the

distributaries and 21 feet in the village water courses. This left 53 feet delivered on to the field out of which 25 feet were wasted in different ways. The efficiency of the canal was then 28 per cent. This statement was made on the basis of the findings of 1882-3. According to P.J. Fagan, Revenue Secretary to the Punjab Government, there was improvement in the subsequent years and the efficiency level was probably 35 per cent. For details see *Report of the Indian Irrigation Commission,* 1901-3, Vol. III, p. 60.

27. *Report on the Factors Affecting Irrigation in the Punjab From Tube Wells* (Labore, 1938), p. 5.
28. See notes by Chief Engineer, Irrigation Works, 6 July 1920 and by Inspector General of Irrigation, 28 July 1917 cited by Imran Ali, *Punjab Under Imperialism* 1885-1947 (Princeton University Press, 1988), p. 176. For the same view see also R.D. Tiwari, *Indian Agriculture* (Bombay, 1943), pp. 238-40; Howard, *Crop-Production in India,* p. 82; Govt. of Punjab (Irrigation Branch Paper No. 24), *The Sale of Water For Irrigation At Volumetric Rates* (Lahore, 1920), p. 1, and *Report of the Royal Commission on Agriculture in India,* Vol. VIII: *Evidence Taken in the Punjab,* pp. 414-15, 444.
29. According to one estimate the cultivators used about 20 per cent more water than would be required to give the same outturn if water were applied with due regard for economy. For details see *Punjab Memorandum to the Royal Commission on Agriculture in India* (Lahore, 1926), pp. 89-90.
30. *Report of the Royal Commission on Agriculture in India,* p. 16.
31. *Final Settlement Report of the Jhang and Gugera Branch Circles of the Lyallpur District* (Lahore, 1925), p. 4.
32. For details of these measures see *Memorandum Submitted to the Royal Commission on Agriculture,* pp. 82-90; *Season and Crop Reports* (annual) and the Irrigation Branch Papers.
33. For details see *Administration Report of the Punjab Pubilc Works Department* (Irrigation Branch), 1930-4, p. 63.
34. Percentages of area affected by waterlogging problem in the different districts of West Punjab were as follows: Lahore 9; Sialkot 41; Gujranwalla 87; Sheikhpura 66; Gujrat 12; Shahpur 13; Mianwali 7; Montgomery 12; Lyallpur 48; Jhang 49; Multan 15 and Muzaffargarh 48. For details see *Agricultural Statistics of the Punjab, Pakistan,* p. 15.
35. *Report of the Indian Cotton Committee,* Vol. 1 (Calcutta, 1919), p. 28. Albert Howard (*op. cit.,* p. 31) observed that over-irrigation itself adversely affected yield rates of crops.
36. *Report of the Royal Commission on Agriculture in India, op. cit.,* p. 522.
37. *Census of India,* 1931, *Punjab, Report,* Vol. XVIII, Part 1, p. 18.
38. Testimony of the Director of Public Health to the Royal Commission on Agriculture, *Report, op. cit.,* p. 538.

CHAPTER 3

IRRIGATION AND CROP PATTERNS

In the last chapter it has been shown that provision for increased water supply led to a vast expansion of cultivated area in the Punjab through breaking-up of virgin lands, expansion of irrigation to areas which were not irrigated in the past, reduction in following and crop failure as well as improved yield rates[1] of crops. Some idea about the increased crop production which was thus possible may be made from data presented in Table 3.1. But it is obvious that irrigation does not increase agriculture production only through these means, but also by promoting transfer of resources to superior or more valuable crops. This explains why inquiry in this latter area is usually regarded as a major aspect, sometimes even the starting point, in formal cost benefit analysis of irrigation projects. Therefore, the direct economic effects of the Punjab irrigation system can not be judged only in terms of the expansion of cropped area impressive though it was, it will be important to examine whether and to what extent increased water supply encouraged or facilitated substitution of inferior crops by superior ones.[2] It is then clear that attention should be directed to two questions: (a) How did irrigation influence crop-combinaton? (b) Were the crops that increased their share of the total irrigated area more valuable or profitable than other crops and, if so, in what sense?

The question whether the farmers in the underdeveloped countries select a crop-mix which would maximize their income within the given institutional and technological constraints has generated an intense debate among the economists. But so far as India is concerned the debate about peasant behaviour is not new—it started much earlier during the British rule, though the question was not posed in this manner. As agricultural production became more market-oriented the nationalist spokesmen, particularly R.C. Dutta, argued that this was not out of the free will of

TABLE 3.1: ACREAGE, YIELD AND OUTPUT OF GROUPS OF CROPS (1906-7–1946-7)

Crops	*Area*			*Yield*			*Output*		
	a	*b*	*c*	*a*	*b*	*c*	*a*	*b*	*c*
All crops	1.39	-0.19	0.43	0.30	-0.05	0.36	1.57	-0.49	0.79
Cereals & Pulses	1.00	-0.21	0.31	-0.01	-0.02	0.11	1.10	-0.46	0.42
Cash crops	2.29	-4.62	1.08	0.86	0.23	1.55	3.21	-82	2.64

Source: Computed from data in the Appendix.

Note: (a) Irrigated; (b) Unirrigated; (c) Total

the cultivators—commercialization was forced, artificial. This was so because the high pitch of revenue demand in cash compelled the cultivators to sell a large portion of the produce of their fields keeping an insufficient stock for their own consumption.[3] On the other hand, members of the colonial bureaucracy argued that it was market forces rather than the pressure of land revenue that were drawing the farmers into the business of production for the market. The commercial crops were more profitable and this economic incentive led them to produce for sale and export, thus making it possible for them to increase per capita income.[4] This debate has continued between two groups of latter-day commentators. Thus, while the historians following neo-classical economic theory have followed the line of the colonial administrators,[5] some of the anti-imperialist (both Marxist and non-Marxist) commentators have supported and extended the contention of the nationalist spokesmen. According to them, given that population was increasing the cultivators would have preferred sowing a larger acreage with food crops (thus reducing cash crop acreage) and/or selling a smaller volume of food crops in the market. But the burden of revenue/rent to be paid in cash and the pressure of the moneylenders led them to divert resources to the production of cash crops (like jute and cotton)/or sell more food crops than could be justified taking into consideration their consumption needs.[6] It is then clear that a closer look at the questions raized earlier can shed light on some of issues in this debate about peasant behaviour.

3.1 Mode of Presentation and Sources of Data

We propose to indicate the impact of irrigation on crop-combination by comparing trends in total irrigated acreage with those in individual crops and groups of crops. For this purpose the following crops have been selected: wheat, gram, barley, maize, jowar, bajra, rice, sugarcane, cotton, tabacco, rabi oilseeds (linseeds, sesamum and mustard) and "fodder crops not otherwise shown". To facilitate the interpretation of the estimated trend rates rice, jowar, bajra, barley and maize have been treated both individually and as well as a group—minor cereals. The same procedure has been followed in the case of the cash crops—cotton, sugarcane and oilseeds. "Fodder crops not otherwise shown" were lumped together and statistics under this category were first reported in 1897. While the cash crops were produced primarily for sale, the other crops were raised partly for sale partly for on-farm consumption.[7]

This mode of analysis is not naturally an ideal one. For irrigation was not the only influence on the production decision of the cultivators. Relative price changes, communication development, demographic and institutional factors are also likely to have had some impact on the selected crop-mix. Moreover, systematic manipulation of water supply by the Hydraulic Regime for production of certain crops, differences in water rates and introduction of new varieties of seeds could also have influenced the performance of different crops. Therefore, wherever possible we have referred to the other relevent factors. With regard to the relative profitability of crops, though statistics on prices and yield rates of crops are available for every district, reliable information on the cost of production is hard to come by. The Punjab Board of Economic Inquiry collected some farm-level data in the 1920s and 1930s. But these were based on such small samples that it will be risky to attempt any firm generalization. Therefore this chapter will be primarily concerned with the impact of an assured water supply on the crop-mix. The pattern that emerges from the available farm-level data on the return per acre of crops is also discussed, but this will not be an attempt to say anything conclusive about the profit-maximising behaviour of the Punjab cultivators.

The relevent statistical data have been drawn from the *Punjab Revenue Administration Reports* (annual) for the period upto 1900-1 and *Season and Crop Reports* (annual) for the subsequent years. As mentioned in the case of the trends in irrigated acreage, there are two elements of non-comparability in the data: one caused by boundary changes and the other by the fact that acreage data recorded before and from 1906-7 were respectively those on gross harvested area and sown area. To overcome this problem trend rates have been estimated separately for the years 1887-8–1900-1 and 1906-7–1946-7. As mentioned earlier, the acreage data are generally believed to be reliable. With regards to yield statistics, there are good reasons to believe that the degree of underestimation in these figures increased over time. However, as it will be extremely difficult to make some idea about the magnitude of this increase no attempt has been made to revise these statistics.

3.2 Estimated Rates of Changes During 1887-1900 and 1906-46

The average annual rates of changes in irrigated and unirrigated acreages under individual crops and two groups of crops are presented in Table 3.2. These indicate that there was a marked disparity between the performance of the two groups of (a) cash crops and (b) cereals and pulses, the irrigated area under the former group increasing much faster. This discrepancy meant that as groups cash crops increased their share of the total area served by public canals, wells and other miscellaneous sources of irrigation from 14 per cent during 1887-8–1895-6 to slightly more than a fifth during 1916-17–1925-6 (Table 3.4). The upward trend lost momentum during the Depression years and it was actually reversed in the last decade which saw an improvement in the relative prices of food crops and the Grow More Food Campaign launched by the Punjab government. But the underlying pattern is clear enough—there was a shift of resources to the cash crops. This trend is underlined by three other developments of this period. Firstly, the decline in the proportion of unirrigated area sown with cash crops was more marked than in cereals and pulses. Secondly, the share of these crops in irrigated area continued to increase even during the Depression years. Thirdly, the percentage of unirrigated area devoted to cereals and pulses declined even though population was increasing in the province at the annual rate of 1.2 per cent during 1901-94. As about the other crops Bagchi has surmised that irrigation "displaced dry crops such as millets or pulses".[8] But the estimated trend rates in gram and the minor cereals taken together do not lend support to his contention. If some displacement may be said to have occurred at all, it was in the rain-fed (*barani*) area.

Thus, performance at the aggregate level clearly indicates that the expansion of

TABLE 3.2: TREND RATES IN ACREAGE AND YIELD

Crops	*Acreage*				*Yield*	
	1887-1900		*1906-46*		*1906-46*	
	a	*b*	*a*	*b*	*a*	*b*
Total	2.41	-3.05	1.39	-0.19	0.30	-0.05
Cereals & Pulses	1.80	-3.66	1.00	-0.21	-0.01	-0.02
Wheat	2.00	-2.23	0.82	-0.09	0.26	0.04
Gram	2.92	-6.15	1.77	-0.13	-0.35	-0.74
Minor Cereals	1.26	-4.45	1.04	-0.09	-0.02	-0.15
Cash Crops	3.46	-4.64	2.29	-1.62	0.86	0.23
Cotton	4.35	-2.81	2.88	-2.57	1.78	1.12
Sugarcane	1.44	-2.74	1.18	0.18	0.75	0.04
Oilseeds	4.24	-5.67	-0.32	-0.97	-0.16	-0.21
Tobacco	1.13	-0.04	0.29	-0.03	-	-
Maize	1.97	0.52	0.70	-0.25	0.15	-0.05
Barley	0.59	-8.65	-0.70	-1.95	-0.32	-0.48
Jowar	-0.84	-7.34	4.16	-0.98	0.20	-0.38
Bajra	3.91	-3.80	3.33	0.77	-0.34	-0.46
Rice	1.56	-5.16	1.72	-0.16	0.20	0.48

Source: Computed on the basis of data in the Appendix.
Notes: (a) irrigated and (b) unirrigated. Trend rates in yields do not refer to all the crops included in this group. This is because yield rates of all the crops were not estimated by the government.

canal-irrigation led to greater emphasis on cash crop production. But the picture appears to be more complicated when we look at the trends in individual crop acreage. Irrigated area under oilseeds (mustard, sesamum and linseed) expanded much faster than total irrigated area during the first fourteen years, but declined in absolute terms during the period 1906-7–1946-7. Extension of area under sugarcane always lagged behind the expansion of water supply facilities. Thus, the fact that as a group cash crop fared better than cereals and pulses is due to the marked expansion of cotton acreage. The pattern was mixed also among the individual cereals and pulses. Area under bajra and gram expanded faster than the spread of irrigation althroughout the six decades under review. On the other hand, irigated area sown with jowar always declined. The fact that the performance of the individual crops differed a great deal is also underlined by the contrast between rice and wheat. Expansion of acreage under neither crop matched the spread of irrigated area during the first 14 years, but the disparity was more marked in the case of rice. The pattern changed during 1906-7–1946-7. For whereas rice area now expanded faster, the disparity between area under wheat cultivation and the spread of irrigation became even more pronounced.

Before we proceed further to analyse the factors behind the performance of the individual crops it will be well worth emphasizing two important aspects of the

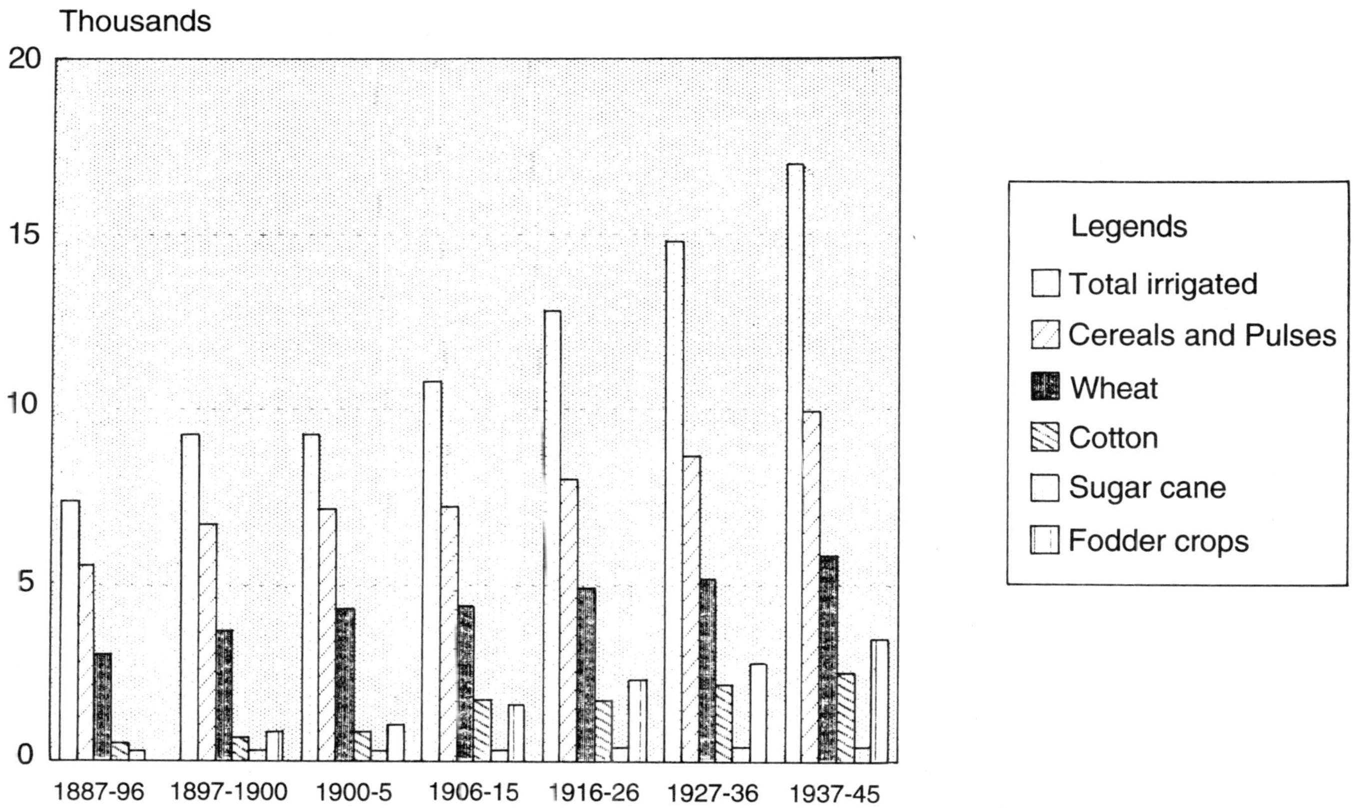

GRAPH 2(A): IRRIGATED AREA UNDER CROPS (IN '000 ACRES)

Source: *Revenue Administration Reports* and *Season and Crop Reports*

agrarian scene of the province which have emerged from our discussion so far. Firstly, after six decades of impressive growth the share of the main cash crops in the total crop-mix remained at about one-fifth. Such a performance does not match the weight which these crops are usually believed to have assumed in India under British rule. This point deserves emphasis also in the sense that cash crop acreage does not adequately reflect the extent of commodity production in Punjab agriculture. Under British rule agricultural production in this province became one of the most market-oriented in the whole of Asia because a foodcrop like wheat also became a major export crop. It may be argued that since yield per acre of some of these (cash) crops was improving faster and crops like sugarcane and cotton fetched a much higher price, annual expansion rate of irrigated area under these crops or their percentage distribution in the total crop-mix fail to reflect their real weight in the agrarian economy. But the picture does not significantly improve when we make allowance for these factors—the output of the main cash crops (weighted by the price current during 1924-5–1928-9) still remains less than one-fourth of the total crop-output in the irrigated tracts of the province (Table 3.3). The picture improves only when allowance is made to the marked expansion of irrigated acreage under fodder crops.

The other feature concerns the trends during the Great Depression-irrigated (as well as total) crop acreage increased[9] even though agricultural prices slumped by about 50 per cent. A fuller analysis of the impact of the Depression of Punjab agriculture is outside the scope of this present study. However, some of the more important factors behind the observed trends may be mentioned. Firstly, cereals and pulses occupied 70 per cent of the total crop acerage of the province and the larger part of the production of these crops was retained at the village level for payments in kind[10] and domestic consumption by all the cultivating classes taken together. Falling prices did not naturally have an adverse impact on this part of production. Secondly, even the small and marginal farmers had certain cash needs, becuase they had to pay revenue and interest charges on money-loans contracted during the years of rising prices. Moreover, a section of the tenants paid cash rent. With the fall in agricultural prices the burden of these "paid-out" cost of cultivation increased, thus making it necessary for the cultivators to sell a larger quantum of their produce. It may be mentioned here that these cutivators mostly depended on family labour for their agricultural operations and labour had little opportunity cost. Thirdly, the other groups of cultivators and the landlords were also under pressure to increase their production. This is because during the years of rising prices they had become used to a certain life-style which involved considerable cash outlay and these cash needs did not decline as much as agricultural prices did. It is worth emphasizing here that the rate of cash wages paid to agricultural labourers declined (though after a time lag) during these years. This implies that the surplus farmers who regularly employed wage labour were not disadvantaged to such an extent that they would refrain from maintaining the level of production at the pre-depression level. Fourthly, land had virtually no alternative use. Thus, several inter-related considerations made it impossible for the cultivating classes and the landlords to reduce the level of agricultural operations even though there was a deep cut in prices.

It may be argued that even if cultivators had to expand their total sown area, use

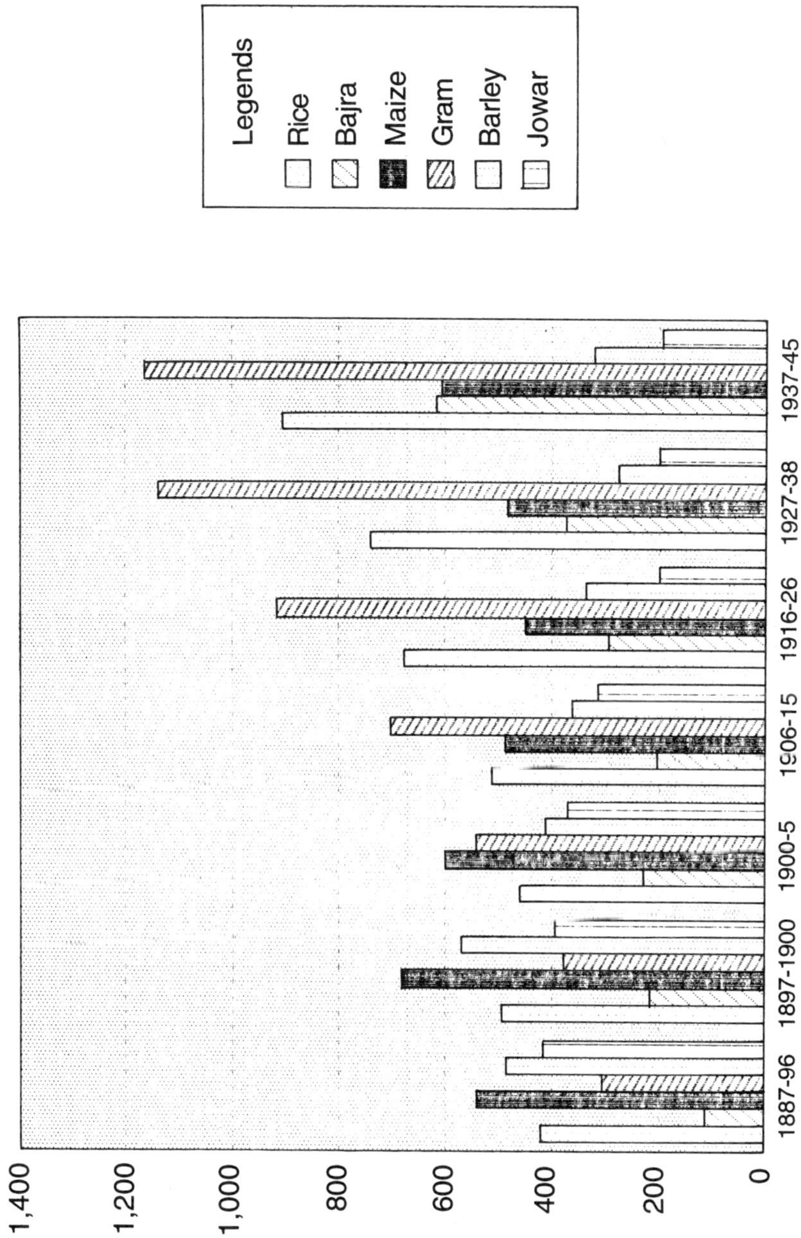

GRAPH 2(B): IRRIGATED AREA UNDER CROPS (IN '000 ACRES)

TABLE 3.3: WEIGHTAGE OF THE OUTPUT OF GROUPS OF CROPS

Year	Cash Crops		Minor Cereals		Foodgrains	
	a	b	a	b	a	b
1907-16	17	8	20	28	83	92
1917-26	21	8	18	28	79	92
1927-36	21	7	17	26	79	93
1937-46	23	8	17	26	77	92

Source: Computed on the basis of data in the Appendix
Notes: (a) irrigated, (b) unirrigated

of canal water which had to be paid for in cash should now have been reduced. But there are several causes which would explain why it did not happen. Firstly, in many parts of the province where the perennial canals were constructed agricultural operations were not possible without artificial water supply if weather conditions were not exceptionally favourable. Secondly, crop yield per acre was much higher in irrigated land and this was more secure in the sense that crop failure rate was much lower. But price (per acre) charged for canal water was moderate, thus implying that a large part of the enhanced yield due to canal irrigation accrued to the irrigation. Thus, the incentive which had, in the past, led to a vast expansion of irrigation to virgin lands and cultivated area served by other sources of water (especially wells) or not irrigated at all was, in a sense, still present during the Depression years.

Thus, to summarize this part of the discussion, the pattern that emerges from the comparison of trend-rates in irrigated acreage under individual and groups of crops are quite complicated. Fodder crops registered the highest rate of expansion, thus increasing its share of total irrigated are much faster than any other groups of crops. Among the cash crops only cotton increased its share, especially during the last three decades. Of the other crops, gram and bajra acreage expanded faster but wheat, the most important food crop, lagged far behind. The only generalization which thus seems to be justified is that increased water supply led to a shift in favour of the major cash crop (cotton), minor food crops and some of the fodder crops.

We may now try to explain these differences in the performance of the different individual crops. The pattern is clear enough in the case of cotton. Provision for quality seeds formed an important aspect of agricultural development in the Punjab (especially in the canal colonies) and the achievement was more pronounced in cotton than in any other crop. The Lancashire mill-owners brought pressure on the Indian government not only to increase the quantity but also the quality of cotton. The government responded by evolving a hybrid (Punjab 4F) from a number of American varieties. The new variety (popularly known as "Amreekana") improved yield per acre and fetched higher price because of its superior quality (the staple of 4F being seven-eighths of an inch as against five-eighths in the case of country or *deshi* cotton).[11] Moreover, being a thirsty crop, the new variety was eminently suitable for the canal-irrigated tracts.[12] But two problem stood in the way of its rapid spread—supply of seeds and marketing facilities. In the circumstances the government

TABLE 3.4: DECENNIAL ACREAGE OF IRRIGATED AREA UNDER DIFFERENT CROPS ('000 ACRES)

Crops	*1906-15*	*1916-26*	*1927-36*	*1937-46*
Cereals & Pulses	7,214(103)	8,024(115)	8,656(125)	10,031(145)
Wheat	4,446(104)	4,912(115)	5,183(121)	5,843(137)
Minor Cereals	1,940(96)	1,917(97)	2,059(100)	2,638(130)
Cash Crops	1,739(97)	2,549(142)	3,180(177)	3,350(187)
Cotton	1,035(111)	1,588(170)	2,163(231)	2,458(263)
Fodder	1,614(128)	2,289(181)	2,794(220)	3,483(275)
Total Irrigated Area	11,109(106)	13,175(125)	14,969(149)	17,269(164)
Total Sown Area	2,8071(99)	29,578(104)	31,356(110)	32,678(115)

Source: Computed on the basis of data in the Appendix.

Notes: Index Numbers are presented in parenthesis (1906-7–1910-11 represent 100) 1906 stands for 1906-7 and so forth.

assumed the responsibility for the supply of a huge and ever-increasing volume of seeds by alloting "seed farm" grants in the colonies. In other words, large chunks of lands were given to private persons on the condition that they would supply a certain quantity of cotton seeds to the Agricultural Department which, in turn, would distribute the seeds to the growers. The biggest of these grants (of 7,220 acres) was allotted to the Manchester-based British Cotton Growers' Association. The marketing problem was solved when after 1915 the Agricultural Department organized successful auctions, thus assuring a premium to the growers. Once adequate supply of seeds and fair price were assured acreage under 4F variety expanded fast—from a few thousand acres in 1910 to 514,000 acres in 1923 and about 70 per cent[13] of the total cotton acreage in 1945. Several other varieties of American cotton (289-F/43, 289/K25, L.S.S. 125 and 199F) were evolved at the Experimental Stations, but 4F remained consistently more popular. The *deshi* cotton had two improved varieties (Mollisoni and Jubilee), but these were more popular in *barni* or rain-fed areas. It is then clear that cotton increased its share of irrigated acreage because the new American varieties were most suitable for land which received regular water supply.

The importance of the trends in yield rates and price as a factors behind the performance of individual crops vis-a-vis the expansion of irrigated area is borne out by the experience of oilseeds and sugarcane. Yield rates of the oilseeds declined and this led to an absolute decline in oilseeds acerage. In the case of sugarcane, improved varieties of seedlings were evolved at the Experimental Station of the Agricultural Department and by 1938-9 more than three-fifths (64 per cent) of the cane acreage were under the new varieties (Co-285, Co-312 and, to a lesser extent, Co-213).[14] At the same time, the process of cane-crushing improved as a result of the introduction of iron-crusher. It was difficult to maintain the purity of the new seedlings. But nonetheless yield rate improved to a considerable extent,[15] thus providing a certain stimulus for the expansion of irrigated sugarcane acreage. Acreage under this crop did not increase faster than total irrigated area because, as elaborated latter, there were several limiting factors. Thus, to conclude this discussion on the factors behind the performance of cash crops, while a decline in oilseeds yield was accompanied by a contraction in sown area, a marked improvement in price and yield rates as a result of the introduction of the new variety saw a more than proportionate expansion of area sown with cotton. Between these two extremes, a moderate improvement in sugarcane yield and price led to some expansion of irrigated acreage under this crop.

To turn to the variations in the trends in irrigated acreage under individual food grains and pulses, the impact of market forces is likely to have been less important, given that the large part of these crops was meant for on-farm consumption. It is also worth mentioning here that often the same plot of land was sown with two of the minor cereals or pulses or these were raized along with wheat or gram. Moreover, the minor cereals and pulses were mostly raized in land cropped twice during the same agricultural season. Thus considered, it is not surprising that there were no significant trends in the yield per acre of minor cereals or any consistent relationship between yield and acreage. As for wheat, it is highly unlikely that the rate of improvement in yield was as low as 0.26 per cent per annum, because by 1945 about 70 per cent of the wheat area was sown with improved varieties of seeds (known as Punjab-11, 8A, etc.) evolved by the Agricultural Department.[16] But two considerations suggest that the new varieties may not have brought about any dramatic improvement. Firstly, progress in the spread of these new varieties was very slow during the first two decades of the present century. Secondly, it was always difficult to maintain the quality of the new seeds of any crop, whether sugarcane, or cotton or wheat.[17] But as wheat was grown on such a large-scale the likelihood is that the problem was more acute in the case of this crop. It would then appear that irrigated area under wheat did not expand very fast during the first there decades (1906-7–1936-7) primarily because yield rate improved rather slowly.

The marked expansion of "fodder crops not otherwise shown" in the irrigated area was made possible by a steady increase in the demand for fodder beyond the point at which this could be met by the fodder contents in crops like wheat, barley, gram, jowar. Firstly, it seems that in the canal tracts livestock population increased faster than in the province as a whole (total provincial increase being of the order of 200,000 between 1905 and 1935).[18] In this connection reference may be made to the fact that in some colonies the government created service tenures under which the grantees had to maintain a certain number of cows, bulls, camels and mares. For example, the whole of Jhelum Colony was reserved for horse-breeding, while in the Lower Bari Doab Colony there were six cattle-breeding farms and the area covered by some of these farms ranged between three and four thousand acres. *Bhusa* (fodder) was also needed by the Military Department for its Dairy Farms as well as for the horses and livestock maintained in the cantonments. Secondly, there was a demand for fodder from industrial firms for the manufacture of cupboards and for export to the United Province. Thirdly, it is also possible that the increased production of fodder crops was looked upon as means for the preservation of soil fertility.[19] Most of the common crops in the province such as wheat, cotton, sugarcane, toria were very exhaustive. Therefore, the inclusion of such leguminous crops in the rotation may have been prompted by a desire to maintain or even improve the productive power of the soil. Incidentally it may be mentioned that fallowing as a proportion of irrigated area was declining in the province.

The indication that changes in crop-mix in the irrigated areas of the Punjab were caused by improvement in yield rates and/or increase in relative price should not surprise us. For it has been known for many years that Punjab farmers responded normally and significantly to relative price and yield changes.[20] However, such a response to market forces does not necessarily preclude the possibility of a significant degree of institutional pressure. For while the large landowners would respond to improvement in relative price/yield out of their free will, it is possible that many cultivations increased cotton production more than they would have normally done because they were under pressure[21] from landlords and moneylenders[22] to do so and, consequently they failed to sow the desired area under wheat or some other foodcrops. Thus, a positive response by the subtenants and small and marginal cultivators in general to relative price and yield changes may actually have been a forced and artificial one.

Should it then be argued that institutional pressure or non-market force did play a certain role in promoting the larger production of cotton or fodder crops in the irrigated tracts of the province? Without going into the details about the debate on the nature of commercialization in Indian agriculture under British rule two points may be emphasized here. Firstly, in 1931-2 about half of the total cultivated area in the province was held by the subtenants–36 per cent by those who paid rent in kind and 12 per cent by those who paid in cash. The *bataidars* (those who leased-out on produce rent) had naturally a stake in the level of productivity of their land, especially because they were paying a part of the cost of production. Consequently, they were led to indirectly intervene in the process of production and one important form of this intervention was to decide the crop-mix of the tenants. As pointed out in a village survey report, "at the beginning of an agricultural year it is usual for the landlord to agree with his tenant as to the crop to be grown in a particular field". We are further told that a "tenant who cares to keep the same land on rent generally follows the advice of the owner in regard to the rotation of crops to be followed".[23]

It is then clear that the landlords were in a position to dictate the production of certain crops. But, as shown later, crops like cotton and sugarcane fetched a higher price. This deserves emphasis because the share-croppers may not have always been unwilling to decide crop-mix according to the desire of the landlords and devote a larger area to these crops than would have been otherwise justified by the need for growing foodcrops for domestic consumption.[24] This was not obviously to earn a profit in the capitalist sense of the term, but to be able to buy a larger quantum of foodgrains for consumption by the members of the family (i.e. a quantum larger than what could have been produced on the same plot of land). Such a production-decision should have appeared reasonable also because in any case many of them had to sell a part of their produce for certain cash needs (payment to the moneylenders and landlords if land was leased in on cash rent as well as to buy certain basic items from the market). Moreover, crops which fetched higher prices were more labour-intensive and the small farmers had an advantage in this respect in as much as the ratio between family labour and cultivated land was more favourable for them.[25] True, such consideration could not have carried much weight for a section among the sharecroppers and their decision to devote a part (or a larger part than they wanted) of the land to the production of a crop like cotton was a dictated one. But it should be borne in mind that their contribution to total cotton output (and indeed to total marketed surplus) was rather small.

Secondly, with regard to the pressure of the merchants and moneylenders, Mridula Mukherjee has argued that there "is no evidence in the Punjab of peasant being tied to the market or to specific traders or merchant moneylenders through advances made on the condition that they produce specified quantities of crops, the price of which was fixed beforehand. This system did not operate even in the most backward parts of the province."[26] She refers to the presence of the *dadan* system in certain jute producing area of Bengal[27] and points out, "In Punjab, however, neither wheat nor cotton production was organized in this manner." Mridula Mukherjee also discusses at length why the pattern was different in this province and according to her the most important factor was that cotton was grown primarily in the agriculturally more prosperous areas such as the canal colonies and parts of central Punjab where water supply was assured. In these areas the number of people who could be forced to accept *dadan* was very few.[28]

There is little scope for disagreement on this point stressed by Mridula Mukherjee, but given the importance of the subject some further elaboration will be in order. Firstly, we would argue that a certain section of the growers did receive *dadan* or advances from the rural merchant-moneylenders or the agents of the export firms (*arhatias*) on the condition that stipulated quantities of the output of a crop would be delivered to them after harvest or the produce would be sold to them at a certain price.[29] But this was not necessarily to dictate the production of the crop in question,[30] the motive was to increase to margin of profit or the interest rate by stipulating a lower price.[31] The cultivators had to accept such terms because the need for credit was urgent and even after the rise of a class of moneylenders from among the agriculturists the *banias* remained "still the most important source of credit".[32] Secondly, the fact that a high proportion of agricultural produce was usually disposed of by the Punjab peasants at the village level rather than in the *mandis*[33] should not be taken as an index of the extent of control exercized by the *banias* or the *arhatias* on production and marketing.[34] The proportion was high because, though all loans were not in the nature of advance, many indebted cultivators disposed of their produce through the village merchants. Moreover, in many cases agricultural produce was sold at the village level even by those who were not in debt.[35] Thus, to conclude this part of the discussion, as far as the overall pattern with regard to the greater emphasis on the production of a crop like cotton is concerned, institutional pressure (i.e. pressure of the landlords and moneylenders) was of limited significance.[36]

3.3 Profitability of Crops

The question that needs to be addressed at this stage is whether the crops that fared better were more profitable and, if so, in what sense. Estimated gross return per acre and proportion of area occupied by certain crops during a decade (1924-5-1933-4)

TABLE 3.5: GROSS VALUE OF PRODUCE PER ACRE OF IRRIGATED CROPS (1924-5–1933-4)

Crops	*% of Total Irrigated Area*	*Gross Value (Rs)*	*Yield per Acre (lbs)*	*Price per Maund (annas)*
Wheat	35	35	901	56
Cotton (A)	6	96	830	145
Cotton (D)	7.5	59	864	125
Sugarcane	2.5	114	1,811	82
Gram	8	27	678	51

Source: Computed from data in the Appendix.
Notes: A = American, D = *Deshi* or indigenous.

are presented in Table 3.5. If we are to generalize on the basis of these data it would appear that sugarcane cultivation was the most profitable. On the other hand, return per acre was one of the lowest in the case of wheat. But whereas wheat occupied one-third of the irrigated area, only 2.5 per cent of the area was devoted to sugarcane cultivation. With regard to cotton, it has usually been credited with the role of "leading sector" in the commercialization of Indian agriculture under British rule, but gross return per acre of the traditional varieties was much lower than that of sugarcane. The comparative picture significantly improves when we look at the American varieties. It may be recalled here that irrigated acreage under these varieties was fast expanding from the early 1920s. But taken together American and *deshi* varieties did not occupy even 20 per cent of the total irrigated area of the province. It will be naturally argued that gross value per acre of a crop can not be taken as an index of its profitability, i.e. the cost of production should also be taken into consideration. We have mentioned earlier that reliable data on the cost of production of different crops are not available from our sources. However, if the farm-level data collected by the Punjab Board of Economic Inquiry are taken for whatever they are worth, it would appear that the difference in the level of cost partly neutralizes relative profitability, but the underlying pattern remains the same. In other words, crops with highest gross returns per acre remain the most profitable ones (Table 3.6).

How can we explain the pattern that emerges? Why did the cultivators refrain from switching to the more profitable crops (sugarcane and cotton) in a more pronounced manner? Calvert a senior I.C.S. officer under Punjab government, tried to examine these questions in the early 1920s. His explanation is simple enough:[37]

> The cultivator has not yet reached the point where he would be able to choose that use for his land which will yield the highest net profit. . . . He grows wheat not because wheat yields him the maximum net profit on his labour and land, but because everyone around him grows wheat because wheat is readily saleable and because with water readily available it is an easy crop to grow.

He admits that the questions whether there was adequate demand for the increased output of a crop, communications were developed in certain areas and highest profit *per acre* could be the only criterion of profitability should also receive due attention.

TABLE 3.6: GROSS AND NET RETURNS PER ACRE ACCORDING TO FARM-LEVEL DATA

Crops	*% of area*	*Yield (maund)*	*Gross Return (Rs)*	*Cost (Rs)*	*Net Return (Rs)*	*Net R. per Unit of Cost*
Wheat	45	19	73	38	35	0.92
Sugarcane	1.4	50	212	139	73	0.53
Cotton	20	8	90	29	61	2.10
Gram	5.8	-	36	20	16	0.80
Fodders	14	-	51	20	38	1.90

Source: Punjab Board of Economic Inquiry. *Studies in the Cost of Production of Crops in the Punjab* (Lahore, 1934).

Note: Figures have been rounded to the nearest maund and Rupees. *Kharif* and *Rabi* Crops have been lumped together. These figures relate to the period 1927-8–1931-2.

Calvert also mentions that in the more progressive tracts "the cultivator is beginning to see that he must live, not on the products of his holding, but on the profits derived from it", and accordingly there was a tendency "to grow a more profitable crop for sale and to buy the needs of the households from the proceeds".[38] But the main thrust of his contention is that "the province seems to have arrived at the stage at which the tendency is to grow that which is most readily saleable".[39] As we argue later, the other considerations mentioned by Calvert are important indeed, but the explanation for the larger production of wheat is not as simple as he would have us believe. Moreover, his analysis does not explain why more cotton was not produced even though this crop was also easily saleable and the peasants had long been acquainted with this crop.

The above questions have also exercised the mind of several economists concerned with the problems of underdeveloped agriculture. Thus, Michael Lipton has suggested that the farmers in the underdeveloped countries face a high degree of risk due to high rainfall variance and of a abnormally severe outcome of crop failure: (a) crippling burden of debt (b) the need to sell land at well below normal market price and (c) attached labour. Therefore, they try to minimize risk by spreading it over several crops.[40] There is no doubt about the validity of such an explanation in a general sense, but as far as the Punjab is concerned this remains inadequate. For water supply was more or less assured in the canal irrigated tracts and the greater part of the land in these districts was held by the large and big landowners. The importance of the latter feature is obvious. For, as Lipton himself points out, risk premium is a decreasing function of assets. According to Bagchi the farmers do not shift their resources to more profitable crops completely because in addition to risk relating to the level of produce, there is the possibility that prices will decline, thus making these crops less profitable or even unprofitable.[41] This seems to have been a more serious problem because such a tendency created a feeling among the cultivators that demand for a certain crop was not flexible enough to absorb the increased supply beyond a certain point. But once again, this does not provide the whole explanation. For the possibility that price would decline was also true in the case of a crop like wheat. Thus, the Punjab Board of Economic Inquiry found that though the price of wheat of a particular year had a positive effect on the acreage and production in the subsequent two years, larger production in a given year had a depressing effect on the price level in the subsequent years.[42]

As a matter of fact there were several other problems specific to the crops in question. The Indian Sugar Committee was of the opinion that though short monsoon period and the extremes of temperature were serious obstacles to sugarcane cultivation, area under sugarcane was likely to expand because lower cost of cultivation compensated for lower yields.[43] But the data generated by the Punjab Board of Economic Inquiry make it abundantly clear that within Punjab itself the cost of cultivation was the highest in the case of this crop. Some idea about the nature of the problem may be had from the fact that so far as labour requirements are concerned sugarcane cutivation was eight times more expensive than cotton cultivation (Table 3.6). This meant that though sugarcane cultivation fetched the highest profit per acre, proft per unit of cost was the lowest in the case of this crops. It further implies that labour days and cash outlay required for growing sugarcane in one acre of land could have yielded a larger profit if these were used for the production of an alternative crop, say cotton or wheat. Secondly, sugarcane could hardly be grown without manuring, but this was not as important for cotton.[44] Thirdly, marketing facilities were not as developed as in cotton or wheat. In this connection it may be recalled that one important factor behind the fast expansion of cotton acreage under HYV after 1915 was the development of marketing facilities under government initiative. It seems that demand factor was an important reason for the lack of government attention to sugarcane, for there was no external market for this crop. The internal demand increased when, following the grant of tariff protection in 1932, the number of sugar factories increased from 7 to 33 in 1933. But the progress was suddenly arrested by the imposition of excise duty in 1934. The additional excise duty imposed in 1937 proved still more burdensome and the number of sugar factories declined[45] to only four in 1939. Fourthly, sugarcane occupies land for nearly one year and this could be an obstacle in the way of normal rotational practices. Given these constraints it is not difficult to explain why sugarcane cultivation did not expand at a much higher rate even though this crop fetched the highest rate of return per acre.

Similarly, there were several adverse factors in the case of cotton. Firstly, the rate of acreage expansion in the early years after the introduction of HYV was lower because, as mentioned earlier, marketing facilities were not yet developed and adequate supply of pure seeds not forthcoming. The rate of expansion was even lower in the last decade of the nineteenth and first decade of twentieth centuries perhaps bacause prior to the introduction of HYV, cotton did not enjoy a market advantage over wheat in terms of gross return per acre. Secondly, cotton cultivation was adversely affected by waterlogging. As pointed out by the Indian Cotton Committee, cotton could not be successfully cultivated where subsoil water table was even temporarily within a few feet of the surface. Therefore, cotton cultivation ceased in the tracts which became waterlogged or in which subsoil water level was within a few feet of the surface.[46] It may be recalled in this connection that in 1946-7 some 2,400,000 acres or 24 per cent of the area under canal-irrigation was affected by

TABLE 3.7: COST OF PRODUCTION ACCORDING TO FARM-LEVEL DATA

Crops	*Manual Rs*	*Labour Days*	*Bullock Rs*	*Labour Days Rs*	*Water Rates*	*Seeds Rs*	*Manures Rs*
Cotton	4.0	12.7	7.4	4.3	6.25	1.0	2.4
Wheat	5.3	20.0	15.0	8.9	5.25	2.4	0.9
Gram	3.5	11.8	7.8	4.6	3.25	1.2	2.0
Sugarcane	33.4	106.9	32.2	29.9	12.00	15.5	15.0
Fodder	2.6	8.0	5.2	2.4	1.75	2.5	5.8

Source: Punjab Board of Economic Inquiry, *Studies in the Cost of Production of Crops in the Punjab* (Lahore, 1934).

Note: *Kharif* and *Rabi* fodder crops have been lumped together.

waterlogging in the 13 districts of Pakistan and out of these 1.3 million acres were being cultivated.[47] Thirdly, it was observed by the contemporaries that the quality of 4F variety of cotton deteriorated after several years use.

Thus, to summarize, increased water supply in the Punjab during 1887-1947 led to a shift in favour of the major cash crop (cotton), three minor foodcrops (rice, bajra and gram) and some of the fodder crops. Sugarcane and wheat acreage also expanded but the expansion rate lagged behind the extension of irrigated area. On the other hand, several crops including oilseeds declined in absolute terms. The decline in the proportion of area under wheat seems to be in conformity with the profit-maximization principle, given that gross return per acre and certain farm-level data on the cost of production suggest that this was one of the least profitable crops. But the pace of decline was marginal and consequently wheat acreage never fell below one-third of the total irrigated area. On the other hand, sugarcane acreage never exceeded 3 per cent even though judged by the same criterion of return per acre this was the most profitable of all crops. One of the most important constraints in the way of a more rapid expansion of sugarcane acreage was that, in the face of only a moderate improvement in yield rate, profit per unit of cost remained low, indeed one of the lowest. Cotton enjoyed a decided advantage in this respect: some improvement in the yield rate on the one hand and lower labour requirements on the other meant that profit unit of the cost of production was the highest (Table 3.7). Thus, the greater emphasis on cotton production should also be viewed as a shift of resources to more profitable crops. But this crop did not occupy even one-fourth of the total irrigated area. In other words, expansion of irrigation facilities can not be said to have led to a specialization in the production of even the second best crop. If there was a specialization at all, it was still in wheat production.

NOTES

1. Here we have also to consider that yield rates of irrigated crops were higher than those of non-irrigated ones. Some idea about the magnitude of these differences may be had from the following instances: Wheat: irrigated 938, unirrigated 551; Gram: irrigated 767, unirrigated 544; Cotton:

irrigated 582, unirrigated 366; Sugarcane: irrigated 1786, unirrigated 1291 (all yield figures are in lbs.).

2. Data generated by the Farm Management Survey in India indicate that if all costs (those actually incurred as well as costs imputed to owned inputs) are taken into consideration, majority of holdings would appear to suffer losses. Accordingly, Krishna Bharadwaj had argued that the profit maximization hypothesis can be saved only by conceding that all farms are in a state of nonviability. It is possible that most of all crops were also produced at loss by the Punjab farmers of our period. But even then exercise like the one being attempted here would be useful in the sense that it would give us some idea as to whether the cultivators were switching over to crops in which the losses were the minimum. For a summary of the findings of Farm Management Surveys see K. Bharadwaj, *Production Conditions in Indian Agriculture* (Cambridge, 1974).
3. R.C. Dutta, *Economic History of India in the Victorian Age,* 1837-1901 (London, 1906), pp. 348-9.
4. While opposing the cotton lobby in Britain for state intervention Sir Charles Wood argued, "My conviction is that an adequate demand evidenced by a rise in price, will produce an adequate supply." An official entrusted with the task of introducing improved varieties of cotton in Berar said, "The first cotton merchant who offered a fraction of an anna more for clean than dirty cotton did more for Wardha cotton than I with all the resources of the Government at my back ever accomplished." Both cited by Akbar Ali Khan, *Some Aspects of Peasant Behaviour in Bengal: A Neo Classical Analysis* (Dhaka, 1982), p. 7.
5. For details see *ibid.*, Peter Harnetly, "Cotton Exports and Indian Agriculture, 1861-70", *Economic History Review,* XXIV (2nd Series), No. 3.
6. Daniel and Alice Thorner, *Land and Labour in India* (Bombay, 1972). The nationalist view has been extended in the sense that according to some commentators the moneylenders also played a role in forcing the commercialization of agricultural produce. The institutional pressures have been described as follows: (a) revenue/rent payment-sale of produce-deficit for consumption-indebtedness-interest payment-sale of produce (b) revenue/rent payment-indebtendess-interest payment-sale of produce and (c) "forcing" production of more cash than otherwise justified.
7. For example, an estimated 60 per cent of total wheat production was retained in the villages for domestic consumption by the cultivators, seed requirements and payments in kind. For details see *Report on the Marketing of Wheat in India* (Bombay, 1937), p. 16.
8. A.K. Bagchi, *The Political Economy of Underdevelopment* (Cambridge, 1983), p. 86 (henceforth referred to as Bagchi 1983); "Foreign Capital and Economic Development in India: A Schematic View" in Kathleen Gough and H.P. Sharma (eds.), *Imperialism and Revolution in South Asia* (New York, 1973), p. 49.
9. The pattern remains unchanged when we compare the crop acreage of the Depression years with the area under cultivation in the years immediately before 1929-30.
10. Deducations for menials' dues, payment in kind to reapers, winnowers and permanent servants amounted, on an average, to about 16 per cent of the total wheat production in Lyallpur and Attock districts. For details see, S. Kartar Singh, *Finance and Marketing of Cultivators' Wheat in the Punjab* (Labore, 1934).
11. H.G. Trevaskis, *The Punjab of To-Day*, Vol. 2 (Lahore, 1932), p. 111. Information on the Government's efforts to evolve new varieties of seeds is available from *Report on the Operations of the Department of Agriculture, Punjab* (annual); Govt.of the Punjab, *A Summary of the Important Results Arrived at or Indicated by the Agricultural Stations and Research Officers in the Punjab,* 1930-1 to 1934-5 (Lahore, 1936); Carl. E. Pray, "The Economics of Agricultural Research in British and Pakistani Punjab, 1905-75" (Unpublished Ph.D. Thesis, University of Pennsylvania, 1978).
12. ". . . any large extension of the cultivation of American cotton in the Punjab is almost entirely a question of an increase in irrigation facilities", *Report of the Indian Cotton Committee,* vol. 1 (Calcutta, 1919), p. 19.
13. William Roberts and S.B.S. Kartar Singh, *A Text-Book of Punjab Agriculture* (Lahore, 1947), p. 437.
14. *Report on the Operations of the Department of Agriculture*, 1938-9, p. 71.
15. Sugarcane yield is also believed to have been underestimated (i.e. it did not reflect the

improvement due to the introduction of new varieties). For details about this argument see Carl Pray, "Accuracy of Agricultural Statistics and Sources of Growth in the Punjab", *Indian Economic and Social History Review,* XXI (2)

16. *Report on the Operations of the Department of Agriculture,* 1938-9, p. 71.
17. According to one opinion about 50 per cent of the wheat crop area consisted of imporved varieties in a high degree of purity, but the remaining area was sown with seeds of a less pure quality. *Ibid.*, p. 77. See also Director of Agriculture, Milne's testimony to the Royal Commission on Agriculture in India (Vol. VIII: *Evidence Taken Punjab,* p. 242); T.G. Kessinger, *Villayatpur,* 1848-1968 (New Delhi, 1979), p. 120.
18. For details on the increase in livestock population see quinquennial *Live-Stock Census Reports*. The question of rise in fodder prices has been dealt with by Malik Amanat Khan, *Rise in the Prices of Bhusha in the Punjab* (Lahore, 1945).
19. Hirashima Shigemochi, *The Structure of Disparity in Underdeveloped Agriculture: A Case Study of the Pakistan Punjab* (Tokyo, 1974), p. 34. Ester Boserup has suggested that the introduction of leguminous crops, together with a rise in the supply of animal manure, helped to bring about a sharp increase in the yields of other crops in the rotation. For details see her, *The Conditions of Agricultural Growth* (London, 1965), pp. 35-9.
20. Raj Krishna, "Farm-Supply Response in India-Pakistan: A Case Study of the Punjab Region", *Economic Journal,* 1963.
21. This point is not given adequate importance by Akbar Ali Khan. He finds that jute acreage in Bengal was price-responsive and, therefore, concludes that there was little or no institutional pressure on the jute-growers.
22. For such a view specifically with regard to Punjab agriculture see Richard G. Fox, *Lions of the Punjab: Culture in the Making* (University of California Press, 1985), p. 7.
23. *An Economic Survey of Gaggar Bhana: A Village in the Amritsar District* (Lahore, 1928), p. 146. Given that the irrigated tracts had a higher incidence of sharecropping tenancy it would follow that the role of the landlords was more pronounced in these areas.
24. This is illustrated by the experience in Bengal. As population growth in a restricted land area led to shrinking plot-size, the typical Bengal peasant could no longer grow enough paddy to feed his family. The consequence was a significant and systematic shift to growing the more profitable jute crop in areas in which it was feasible in ecological terms. This made possible greater earnings in the limited holding and the purchase of food from the market. In fact, peasants continued production of jute despite official propaganda to reduce the area under its cultivation. For details see Sugata Bose, *Agrarian Bengal: Economy, Social Structure and Politics 1919-47* (Cambridge, 1986).
25. With regard to the cropping pattern in West Bengal and Madhya Pradesh Krishna Bharadwaj (*op. cit.*, pp. 69-76) found that in the small firms the proportion of area under cereals was less and that under cash crops like cotton, jute and oilseeds was more than on other holdings. From this she concludes: "It is evident that while the very small farmers do not directly grow the subsistence crops, their production activity is geared to providing subsistence for which they need to raise cash crops."
26. Mridula Mukherjee, "Commercialization and Agrarian Change in Pre-Independence Punjab" in K.N. Raj *et al.* (eds.), *Essays on the Commercialization of India Agriculture* (Delhi, 1985), p. 61.
27. We find it difficult to accept this contention. For several reports indicate that even in the most important jute-producing districts (Dhaka, Mymensingh, Rangpur) the system of advance prevailed only to a limited extent. For details see, Government of Bengal, *Marketing of Agricultural Produce in Bengal* (Calcutta, 1926) and H.C. Kerr, *Report on the Cultivation and Trade in Jute in Bengal and of Indian Fibres for the Manufacture of Paper* (Calcutta, 1974).
28. *Ibid.*, p. 62.
29. Neeladri Bhattacharya, "Lenders and Debtors: Punjab Countryside 1880-1940", *Studies in History,* 1, 2 (1985), p. 313.
30. Neeladri seems to take this position when he asserts "the system of advances was far different from those which operated in the case of opium, indigo and jute production in Bengal or sugarcane in Uttar Pradesh".
31. Neeladri (*ibid.*, p. 307) takes the same view when he says that since the rural traders could not

always determine the price at which they sold their stocks to wholesalers in *mandis*, they attempted to exercise their control at the points of supply in the village. But he takes a different position on p. 313 where he argues that the system of advances functioned primarily as a mechanism for ensuring regular supplies and not for regulating prices. It became a system of forward purchase. As a matter of fact both the systems may have prevailed.

32. *Report of the Punjab Provincial Banking Enquiry Committee,* Vol.1 (Lahore, 1930), p. 22.
33. From the data collected by the Banking Enquiry Committee it appears that the cultivators carried 50 per cent, village shopkeepers 40 per cent and agents of *arhatias* and visiting dealers 5 per cent of the marketed produce of cotton to the *mandis.* For details see *ibid.*, Vol. 2, p. 491.
34. It is necessary to emphasize this point because the high proportion of disposal of cotton produce might give the opposite impression.
35. However, though sale within the villages was more pervasive than advances, the cultivators did not necessarily stand to lose very much from such disposal. For disposal in the market involved cost of carting and payment for different charges. In the case of wheat the differences between market and village prices were found to vary betwen four to little over six annas per maund. For details see Kartar Singh, *op. cit.,* p. 19.
36. One may take the same position on the wider question of the "forced" commercialization of agricultural produce. For though cash obligations to the state/landlords and moneylenders as well as other needs compelled many small marginal cultivators to resort to distress sale, such sales constituted a small part of the total volume sold in the market.
37. H. Calvert, *The Wealth and Welfare of the Punjab* (Lahore, 1922), p. 12.
38. *Ibid.*, p. 48.
39. *Ibid.*, p. 12.
40. Michael Lipton, "The Theory of Optimising Peasant", *The Journal of Development Studies,* Vol. 4, No. 3 (1968).
41. Bagchi (1983), p. 174.
42. Punjab Board of Economic Inquiry, *Some Factors Affecting the Price of Wheat in the Punjab* (Lahore, 1937), pp. 68-9.
43. *Report of the Indian Sugar Committee* (Simla, 1920), p. 99.
44. The supply of farmyard manure was limited and this was often a limiting factor in the spread of cane cultivation. Roberts and Singh, *A Text-Book of Punjab Agriculture* (Lahore, 1947), p. 312.
45. B.S. Saini, *The Social and Economic History of the Punjab* 1901-39 (New Delhi, 1975), p. 270.
46. *Report of the Indian Cotton Committee,* Vol. 1, (Calcutta, 1919), p. 28. A related problem was emphasized in the *Report on the Operations of the Department of Agriculture* (1940-1, p. 108): ". . . it is not possible to entirely substitute American varieties for *deshi*, as practical experience has shown conclusively that in the canal colonies the quality and nature of the soil is not uniform and there are areas on which it would be merely courting disaster to plant American cotton".
47. Punjab Board of Economic Inquiry, *Agricultural Statistics of the Punjab, Pakistan,* 1901-2–1946-7 (Lahore, 1950), pp. 14-15.

CHAPTER 4

IRRIGATION AND INCIDENCE OF SHARECROPPING TENANCY

MEMBERS OF the colonial bureaucracy often described Punjab as a province of peasant proprietors[1] and the same view has been taken by several commentators of the post-1947 period.[2] There is no denying the fact that such a description was valid till about the close of the nineteenth century in as much as the larger part of the land was directly cultivated by the proprietors themselves.[3] Moreover, the peasant proprietors constituted the largest single group among the agricultural population.[4] But as the trend towards the expanstion of cultivation by tenants which had set in after the annexation of the province by the British continued, the pattern changed very soon, and at the turn of the present century the tenants came to occupy the larger part (53 per cent) of the cultivated area (Table 4.1).[5] Area cultivated by the tenants increased by another 4.3 million acres during the next three and a half decades and this was almost entirely due to the expansion of area cutivated on sharecropping (*batai*) system.[6] Such area must have increased in the subsequent years, but the relevant data were not published by the government. Incidentally, the contemporary observers of the agrarian scene of province were unanimously of the view that cultivation by subtenants meant inefficiency in production.[7] Why did, then, Punjab agrarian society develop "semi-feudal"[8] characteristic on such a scale, reportedly entailing economic loss in terms of reduced productivity. Did the development of the irrigation infrastructure and the colonization programme that accompanied it play a role in this respect? Is it also possible that irrigation made a difference in respect of the level of rent? This chapter is an attempt to shed some light on these questions.

However, to begin with, some idea may be made about the basic characteristics of the different forms of tenancy prevailing in the province. Briefly speaking, there were two categories of tenants: occupancy tenants and substenants or tenants-at-will. There was a class of tenants who paid a nominal rent or were free from rent, but they held only a small part of the total land. The Punjab Tenancy Act (1887) defined occupancy tenants as those who, for two generations, had paid neither rent nor services to the proprietors, but only their share of the government assessment. So long fixed rent was paid, the occupation of land by these tenants could not be challenged. Rent payable by these tenants (mostly in cash, but to an extent also in kind) could not be enhanced except through a decree of the court of law. On the death of an occupancy tenant his tenancy right passed (a) to his male lineal descendants (b) failing them, to his widow for life or until remarriage and (c) failing both, to agnates or male collateral relatives. On the failure of legal heirs the tenancy reverted to the landlords. Next to the proprietors the occupancy tenants had a pre-emptive claim to purchase

TABLE 4.1: STATISTICS ON TENANCY, ETC.
('000 ACRES)

Years	*Total Cultivated Area*	*Area held by*				
		Owners	*Tenants Paying Rent*	*Occupancy Tenants*	*Subtenants*	
					Cash	*Kind*
1901-2	25,728	11,807(46)	1,363(53)	2,642(10)	3,605(14)	7,384(29)
1906-7	27,231	12,897(47)	14,026(52)	2,368(9)	4,823(18)	8,074(30)
1911-12	27,922	12,830(46)	14,813(53)	2,467(9)	3,401(12)	8,945(32)
1916-17	28,034	12,889(46)	14,833(53)	2,383(9)	3,566(13)	8,884(32)
1921-2	28,820	12,830(45)	15,654(54)	2,409(8)	3,683(13)	9,562(30)
1926-7	29,788	12,902(43)	16,468(53)	2,456(8)	4,136(14)	9,876(33)
1931-2	29,913	1,281(43)	16,634(56)	2,235(7)	3,728(12)	10,671(36)
1936-7	31,042	12,908(42)	17,644(57)	—	—	—

Source: *Land Revenue Administration Report* (annual). Figures relating to 1936-7 have been drawn from B.S. Saini, *The Social and Economic History of the Punjab, 1901-1939* (Delhi, 1975). Statistics on area held by occupancy tenants and sharecroppers in 1936-7 have not been mentioned by him. Figures in the parenthesis refer to the percentage distribution. About one per cent of the land was held by tenants free of rent.

any land offered for sale in the village in which their holdings were situated. But two clauses of the Act protected the interest of the landlords. Firstly, unlike in Bengal, no right of occupancy could be acquired by a tenant merely by lapse of time. Secondly, eviction of occupancy tenants was allowed on payment of compensation.[9]

The subtenant or a tenant-at-will paying rent in kind or cash was an annual tenant as his rent was determined by agreement between him and the zamindar every year. He had practically no security of tenure. The sharecropping or *batai* system may best be understood by comparing it with the *barga* system of Bengal. Firstly, unlike the Bengal landowners, the *bataidars* (i.e. the landowners) of Punjab were responsible for paying a certain part of the cost of production—payment to the village menials (*Kamins*), Government dues, water rates, seeds, etc.[10] Secondly, while in Bengal the landowners' share of the produce was 50 per cent, there were considerable intra and inter-regional variations in the proportion of produce (from one-fourth to one half or even higher) appropriated by the Punjabi landowners and landlords.[11] These variations were due to different factors—the relative bargaining power of the two parties, nature of crops grown and, above all, the *bataidars*' share in the cost of production. When the *batai* was one-fourth the tenant had to pay the land revenue, water rate and provide his own seed; when the *batai* was one-third the landlord paid part of the revenue and half of the seed. But if the *batai* was half, the tenants-at-will paid half of the revenue and half of the *abiana* or no revenue and the entire *abiana* (water rates).[12] The tenants also rendered certain personal services to the landlords—threshing their share of the produce, harvesting their crops, clearing out the silt in the water course and digging new wells. Moreover, their plough cattle were regularly used for ploughing their landlords' land.[13] It will be naturally difficult to say how common these practices were. For there are also evidence to indicate that the substenants were under no obligation to render such services.[14] But in all probability in many cases the *bataidars* insisted on personal services. This is suggested by the fact that a large part of the tenants-at-will belonged to the menial classes and though the *sepidari* system was in a process of dissolution,[15] many of the menials (*kamins*) were tied to particular landowners.

4.1 The Hypothesis

Now, to turn to the main question, in one respect the expectation would be that the proportion of land cultivated by the subtenants was lower in the Punjab than in any other province. As mentioned earlier, the expansion of irrigation facilities led to an addition of several million acres to the cultivated area in the province over a period of six decades (1887-1946). This means that crop acreage expanded faster than in other or most other provinces.[16] At the same time the proportion of people who earned their livelihood from agriculture was significantly lower in the Punjab—59 per cent as against a British Indian average of 69 per cent.[17] Given these favourable factors the vast majority of families in the Punjab rural areas could have owned an agricultural holding large enough to provide employment to their family members. Thus, ideally speaking, one would presume that there was little scope or need for cultivation by tenants in the province, or at least in the canal-irrigated tracts. If there were tenants at all, it would further be expected, they would be in a strong bargaining position, thus precluding the possibility of a high incidence of *batai* cultivation in which the landlords claimed, in many cases, 50 per cent or even more of the produce as rent. The same expectation is also created by the fact that agricultural production in the irrigated tracts, especially in the canal colonies, was more market-oriented than anywhere else in the province. For the transformation in the form of rent (from labour to kind and from the latter to cash) is often considered to be a function of the development of money relations.[18]

But such an expectation does not gain plausibility when we consider certain other factors—land grant policy of the Hydraulic Regime and labour—intensity of cultivation in the irrigated areas. On the other hand, if proper weight is given to these factors and certain aspects of the socio-economic life of the province, one would be inclined to believe that canal construction created conditions in which sharecropping tenancy was bound to flourish on a large scale. Among these factors the specific manner in which the government disposed of the colony lands seems to have been the most important. Indeed, this may be said to hold the real key to the question as to why Punjab provides such an example of a high incidence of sharecropping tenancy under conditions of favourable man/land ratio and a marked degree of commodity production in the agricultural sector.

We have already referred to the disposal of land in the canal colonies among the different social groups in the province. Two other aspects of land grant policy need to be emphasized in the present context: (a) size of grants and (b) the virtual exclusion of the landless section of the rural population, especially the village menials (*Kamins*) from the colony lands. With regard to (a), the peasant (*abadkar*) grants accounted for some three million acres. In the Chenab Colony the size of these grants varied from half a square to two square (one square = 27.7 acres). In the Jhelum Colony the minimum peasant horse-breeding grants was 55.5 acres in size and during 1902-6 some 200,000 acres were allotted to them. Actually the peasants received land grants in all the colonies and their share never fell below 50 per cent in any colony and in some cases exceeded 80 per cent. The size of grants to the landlords (who were given land in all the colonies under different names) is not always known. However, some idea may be made from the following instances. In the Chenab Colony the Yeoman and Capitalist grants ranged from two to five and from six to twenty squares respectively and together these grants accounted for 15 per cent of the total allotted land. The horse-breeding Yeoman grantees in the Jhelum Colony who were obliged to maintain 5 to 15 mares each were allotted 1.5 square per mare. The landed gentry grants allotted in almost every colony were 150 to 350 acres in size and a total of 2,300 rectangles were distributed among the members of this group. Incidentally, the landlords also participated in the auction market. Data on the size of grants to the members of bureaucracy, traders, contractors, lawyers, doctors and military personnel are hard to come by. But it can be safely assumed that these were of several squares and rectangles (i.e. at least more than 50 acres). It may be mentioned here that the military grantees received some half a million acres in different colonies. Finally, in the Nili Bar Colony the government rented out much land in temporary leases and the lessees held from 500 to 2,000 acres each.

But while other social groups (peasants, landlords, retired civil servants, lawyers, doctors, etc.) received land grants the government systematically decided against allotments to the landless rural population, whether agricultural labourers, subtenants or caste servitors. For, it was considered to be a "sound principle that in making selection of grants of Government waste, tenants, labourers and other landless men should not as rule be chosen, as their selection involves the aggravation of the difficulty . . . of agricultural labourers".[19] The only exception was the Lower Bari Doab Colony where some 7,000 acres were allotted to the "depressed classes". But two points deserve emphasis in this connection. Firstly, this area was distributed among the converts of the foreign missionaries and the nominees of different religious organizations. Secondly, the size of the grants did not exceed 12.5 acres. Incidentally, it may also be mentioned that the members of the landless class were too poor to buy land in auction market.

The above was only a brief reference to some of the more important aspects of the land grant policy of the British government, but two points are abundantly clear. Firstly, members of the landless stratum of the rural society were deliberately left out of the colony grants. Secondly, the size of "owned" holdings was generally going to be much larger in the colonies than in the older districts. We can make some idea about the contrast from the following data: (a) A statement prepared by the Punjab Board of Economic Inquiry shows that while 55.548 grants were under one rectangle or square (27.5 acres per rectangle) in size, 77,803 measured one rectangle or square and above.[20] But at the provincial level the average size of ownership holding was seven or eight acres only.[21] (b) It appears from the same survey that in certain *Tahsils* of the colony districts area covered by owners holding of 25 acres and above accounted for 67 per cent of the total cutivated area as against a provincial average of 46 per cent (Table 4.2). (c) Surveys conducted by Calvert under the auspices of the Punjab Board of Economic Inquiry revealed that in the province as a whole holdings of 58 per cent of the owners did not exceed five acres in size (Table 4.3), but in the major colony districts (Jhang, Lahore, Montgomery, Lyallpur, Gujranwalla, Sheikhpura, Shahpur) 43 per cent of the owners had such holding.[22] The contrast between the colony and most of the older districts can be seen even more clearly from the data on the distribution of owned (and cultivated) land among the holding of different sizes as generated by Mridula Mukherjee on the basis of Calvert's findings (Table 4.3).[23] Thus, overall, the colony districts had a larger concentration of such large holdings in which self-cultivation (i.e. cultivation without the help of hired labour on a substantial scale) was not going to be possible. Indeed, many holdings were so large that if the grantees were to assume operational responsibility (i.e. launch large-scale farming) it would have been necessary to use labour-saving implements like tractors.

Such a state of affairs creates a strong presumption that an active lease market would be a marked feature of the agrarian scene of the province. In other words, while there would be a considerable demand for leasing-in land from the landless section of the rural populations (i.e. *Kamins* or those who performed functions as prescribed by their caste status and also worked as agricultural labourers) and the marginal or even medium owners in the order districts, such lands would be readily

TABLE 4.2: PERCENTAGE DISTRIBUTION OF LAND UNDER VARIOUS SIZES OF HOLDINGS IN SOME TAHSILS

Name of Tahsil	*Under 5 Acres*	*5 and under 10 Acres*	*10 and under 15 Acres*	*15 and under 25 Acres*	*25 and under 50 Acres*	*50 Acres and over*
Sargoda	3.4	7.2	8.7	12.9	19.5	48.3
Khanewal	6.3	7.6	9.6	12.1	23.3	41.1
Montgomery	10.2	12.2	7.2	20.9	18.5	31.0
Okara	11.1	18.2	11.2	20.7	18.3	20.5
Dipalpur	3.4	7.0	6.0	18.7	28.8	37.0
Jhang	8.0	10.0	6.2	10.9	18.2	46.6
Chiniot	5.7	8.1	6.8	12.6	23.3	43.4
Lyallpur District	2.1	7.5	13.6	6.5	50.8	20.0
Total (a)	4.0	8.4	10.4	10.7	35.6	30.9
Provincial average (b)	11.0	15.1	11.5	15.2	20.4	25.7

Source: Computed on the basis of data in H. Calvert, *Size and Distribution of Agricultural Holdings in the Punjab* (Lahore, 1925), p. 35.

Note: (a) Percentage of area in these Tahsils taken together;
(b) These figures are drawn from Table 7.3.

available in the canal-irrigated tracts, especially the colony districts. The *kamins* would compete in the lease market because tenancy would confer on them a certain degree of social respectability, i.e. *izzat*.[24] Incidentally, the *kamins* constituted the lowest strata of the Punjab rural society. Indeed they were considered so low that even when hard-pressed by hunger during a bad season, the landowners were reluctant to work with them in road or canal construction.[25] Secondly, the cultivation of such land would entitle them to an income more stable and regular as well as probably higher than the income earned by them as agricultural labourers[26] and/or in their traditional professions. This change in status may not be complete in the sense that all the *kamin*-tenants would not altogether abandon their traditional occupation and/or discontinue "selling" their labour power to the large farmers. But overall, there would be a marked tendency among the *kamins* to lease-in land in the colonies.[27] The *Sepidari* system was in the process of dissolution, because under the changed circumstances the traditional occupations failed to provide employment throughout the year and did not fetch sufficient income for subsistence. But alternative employment opportunities were not being created through industrial development. The expectation that the small and marginal proprietors of the older districts would migrate to the canal colonies to work as subtenants is based on two considerations. Firstly, there was the prospect of leasing-in of a larger holding, or at least holding of a size on which they would be able to expend the available family labour and optimize the use of their indivisible instruments of production. Some idea about the nature of the need for larger holding may be had from the fact that at the provincial leval over 70 per cent of the ownership holdings were of the size from one to ten acres (Table 4.3), but according to one estimate a pair of bullock could cultivate about 15 acres.[28] Secondly, as mentioned earlier, yield rates in the irrigated lands was not only higher but also more secure (given that the risk of harvest failure was much lower). Thus, there are grounds to presume that the landless population and the small peasant-proprietors would migrate to the canal colonies in large numbers to cultivate land as subtenants. It is true that they would be obliged to put in additional labour and capital to prepare the soil as well as build a house and prepare the water courses, but the considerations mentioned above would outweigh these difficulties.

How would the large and big grantees in the canal colonies respond to this situation? To continue the discussion at *a priori* level, it seems that there would be a marked tendency to leaseout a part or whole of the holdings to the subtenants. A section of these grantees in the colony districts (as also some big landowners in other parts of the province) would launch direct cultivation with the help of hired labour and land cultivated in this manner will constitute a significant part of the total sown area. But cultivation with the help of sharecroppers will be the dominant pattern. In other words, the presence of a large body of landless or near landless people ready to lease-in land at a high price is likely to rule out transition to capitalist farming. It may further be suggested that the tendency to employ sharecroppers would not remain confined to the non-agricultural classes among the allottees or auction-purchasers, (a) larger grantees (Yeoman and Capitalist) among the agricultural classes and, in some cases, also the (b) holders of *abadkar* (peasant) grants would employ tenants-at-will. The expectation that the Yeoman and Capitalist grantees would respond in this manner deserve special emphasis. For the Hydraulic Regime had expected that being the "most intelligent and enterprising men in the agricultural community" and endowed with "good credit and resources"[29] they would spearhead agricultural progress. Some peasant grantees are also expected to lease out a part of their land because even though the British government had declared that "it was essential to preserve the tradition of the Punjab as a country of peasant farmers",[30] in many cases their grants were of a size that these could not be self-cultivated. In other words, with the existing level of technology these could be operated only with the employment of hired labour on a large scale. Peasant grantees who would employ hired labour only to supplement family labour or not at all would prefer leasing out their surplus land to subtenants.

So far it has been argued that the land grant policy of the colonial government created conditions in which sharecropping tenancy is likely to flourish on a large scale. Thus, the fact that irrigation facilities were expanding had figured in the discussion only indirectly. But actually irrigation is likely to play direct causative role as well. This possibility is suggested by the fact that agricultural operations were more labour-intensive under irrigated than in rain-fed (*barani*) conditions.[31] Firstly, the task of artificially watering land meant the expenditure of additional labour time. This was particularly the case when water had to be lifted from well even with the help of Persian Wheel. Secondly, more labour-intensive crops like cotton occupied a larger part of the cropped area in the irrigated tracts. Finally, a certain part of the irrigated land was cropped twice in the same agriculture year. Thus, the cost of labour would be higher in the case of production in land served by canals or wells. This should

TABLE 4.3: PERCENTAGE DISTRIBUTION OF (A) OWNERSHIP HOLDINGS (B) OWNED AREA AND (C) CULTIVATED AREA

Districts	*Under 10 Acres*			*10 & under 20 Acres*			*20 & under 25 Acres*			*25 & under 50 Acres*			*50 Acres & above*		
	a	*b*	*c*	*a*	*b*	*c*	*a*	*b*	*c*	*a*	*b*	*c*	*a*	*b*	*c*
Hissar	57	16	23	21	20	29	5	8	12	10	26	28	6	30	8
Rohtak	71	32	44	19	30	37	4	9	8	5	19	10	1	10	—
Gurgaon	78	38	53	14	25	29	3	7	6	4	17	9	2	14	1
Karnal	78	36	41	15	26	30	2	6	8	4	17	16	2	15	4
Ambala	84	39	55	10	22	33	2	7	4	2	10	5	2	22	—
Jabgra	94	65	78	5	15	5	1	4	+1	—	5	—	—	—	—
Hoshiarpur	91	53	74	6	17	11	1	5	1	2	12	1	1	—	
Jullunder	91	60	66	7	22	23	1	5	3	1	7	3	—	4	—
Ludhiana	80	44	44	15	28	35	2	7	10	3	14	8	1	6	1
Ferozepore	66	20	26	17	17	28	3	5	12	6	17	22	8	41	11
Lahore	66	20	27	17	19	36	6	11	14	8	25	20	4	24	2
Amritsar	74	44	50	11	25	30	2	7	8	3	15	8	1	8	1
Gurdaspur	79	35	52	13	24	38	3	8	5	4	17	3	2	16	—
Sialkot	73	46	62	14	30	25	1	3	4	2	14	4	—	5	—
Gujranwalla	60	17	37	20	20	36	6	9	10	8	22	12	6	32	3
Sheikhpura	76	22	26	13	19	38	3	6	9	5	18	22	4	33	3
Gujrat	84	41	55	10	24	30	2	8	6	3	16	6	1	10	—
Shahpur	65	17	23	15	15	29	4	7	10	8	21	28	8	41	9

TABLE 4.3 *Continued*

	a	*b*	*c*	*a*	*b*	*c*	*a*	*b*	*c*	*a*	*b*	*c*	*a*	*b*	*c*
Jhelum	84	43	48	11	25	34	2	7	7	2	12	7	1	12	—
R.Pindi	91	44	63	6	20	28	1	5	1	1	10	2	—	6	—
Attock	63	20	29	19	23	32	5	9	12	8	25	23	4	24	3
Mianwali	75	21	26	12	18	25	3	7	10	6	22	23	4	31	15
Montgomery	64	17	26	14	17	36	6	10	15	9	25	21	6	31	1
Lyallpur	35	12	11	27	16	25	4	5	6	29	36	27	6	31	31
Jhang	67	14	23	12	11	37	3	4	13	8	19	23	11	52	4
Multan	61	13	17	14	12	29	4	6	12	9	20	29	12	50	13
Muzaffargarh	86	32	49	7	16	27	2	5	7	3	16	12	3	30	3
D.G.Khan	80	22	46	8	14	27	2	5	8	4	17	13	5	41	3
Punjab	76	26	34	13	20	30	3		7·9	5	20	19	3	26	8

Source: For (b) and (c) Mridula Mukherjee, "Commercialization and Agrarian Change in Pre-Independence Punjab" in K.N. Raj *et al.* (eds.), *Essays on the Commercialization of Indian Agriculture* (Delhi, 1985) Table 4. For (a) H. Calvert, *The Size and Distributions of Agricultural Holdings in the Punjab* (Lahore, 1925) and *The Size and Distribution of Cultivators' Holdings in the Punjab* (Lahore, 1928) Mridula's camputations are based on the findings of Calvert.

be another reason for which the large grantees or owners of irrigated holdings will not at all launch cultivation with hired labour or employ such labour only upto a certain point.

To summarize that part of the discussion, it may then be suggested that the spread of irrigation would lead to a higher incidence of cultivation by subtenants because *nahri* and *chahi* cultivation was more labour-intensive and there was a higher concentration of larger holdings in the colony districts. This is by no means to suggest that subtenancy on produce and cash rent would not prevail in the rain-fed and other irrigated areas in the older districts. Indeed, a variety of factors—presence of landlords with large holdings,[32] migration of peasants from these districts to the colonies either as (land) grantees or in search of employment as subtenants,[33] ownership of land by widows or orphans or old peasants without family labour,[34] or those who were basically non-agriculturists but resident in the village (i.e. merchant-moneylenders)[35] or had taken up employment in urban areas or gone abroad or joined the army,[36] exchange of land through lease transactions to consolidate operational holdings of cultivating owners[37]—would cause subtenancy to prevail also in these areas. Moreover, such subtenants would include both *kamins*, as well as those who were themselves landowners or occupancy tenants. The simple contention is that the incidence of cultivation by subtenants would be higher in the colony districts and of the two types of subtenants—those paying rent in cash and kind—area operated by the latter group would constitute the much larger part of the cultivated area. With regard to well-lands, it has already been mentioned that labour-intensity was even higher in these areas. This means that, other conditions being the same, incidence of sharecropping tenancy would also be higher in these tracts. Overall, we expect a strong positive correlation between the proportion of area operated on sharecropping basis in the different districts of the province on the one hand and (a) the percentage of irrigated area and (b) area which belonged to those who owned 25 acres or more on the other.

4.2 Data on Tenancy and Test of Hypothesis

Data on the proportion of area cultivated by all classes of subtenants (i.e. those paying rent in kind and cash), sharecroppers, canal irrigated and total irrigated area and families with cultivated holdings of 25 acres and above are presented in Table 4.4. A glance at these statistics reveals that there were 13 districts with irrigated area higher than the provincial average and out of these as many as eight districts had also a higher incidence of sharecropping tenancy. Out of the other five districts the percentage of sharecropped area in Lahore and Ferozepore was similar to the provincial acerage. In Jullunder district where canal irrigation was virtually absent more than nine-tenth of the ownership holdings were less than 15 acres in size. Lower incidence of sharecropping tenancy in Lyallpur, a colony district, was mainly due to the fact that 80 per cent of the land was distributed as peasant (*abadkar*) grants. On the other hand, districts which with smaller proportion of irrigated area, had higher percentages of land operated by sharecroppers numbered only three (Jhelum, Attock and Mianwali). As expected from these statistics, we get a significant degree of positive correlation between the percentages of area leased out on *batai* terms on the one hand and the

TABLE 4.4: DISTRICTWISE STATISTICS ON TENANCY, LANDHOLDINGS, ETC.

Districts	*% of Area Held by Subtenants*		*% of Area Irrigated by*		*% of Ownership Holdings With*	
	Kind Rent	*Kind + Cash Rent*	*All Sources*	*Canals*	*More Than 25 Acres*	*Less Than 15 Acres*
Hissar	19	46	12	11	16	70
Rohtak	10	26	18	14	6	83
Gurgaon	9	34	12	4	5	88
Karnal	18	30	28	16	6	88
Ambala	23	31	04	—	4	91
Kangra	20	23	23	—	1	98
Hoshiarpur	15	26	10	02	2	95
Jullunder	19	31	52	—	1	96
Ludhiana	19	35	39	11	3	90
Ferozepore	31	41	50	34	14	77
Lahore	32	44	82	63	12	77
Amritsar	21	36	72	38	3	91
Gurdaspur	25	38	27	08	5	88
Sialkot	31	40	43	02	3	93
Gujranwalla	43	58	73	69	14	73
Sheikhpura	43	63	93	73	9	84
Gujrat	33	39	45	31	4	90

TABLE 4.4 *Continued*

	Kind Rent	*Kind + Cash Rent*	*All Sources*	*Canals*	*More Than 25 Acres*	*Less Than 15 Acres*
Shahpur	46	55	69	55	16	75
Jhelum	35	36	16	—	3	92
R. Pindi	23	26	02	—	2	95
Attock	49	51	07	—	12	77
Mianwali	36	61	08	02	10	85
Montgomery	78	83	93	70	14	71
Lyallpur	26	43	99	94	35	58
Jhang	56	66	86	14	19	74
Multan	62	70	90	76	21	69
Muzaffargarh	47	50	76	52	6	90
D.G.Khan	95	99	48	37	9	86
Punjab	33	45	46	31	8	85

Source: Data on tenancy have been computed from *Punjab Land Revenue Administration Report,* 1921-2 and those on irrigated area from *Season and Crop Reports*, 1920-1, 1921-2 and 1922-3. Land ownership statistics are obtained from H. Calvert, *The Size and Distribution of Agricultural Holdings in the Punjab* (Lahore, 1925).

proportions of (a) total irrigated area (b) canal lands and (c) large holdings on the other (Table 4.5). A stronger evidence in support of our contention about the importance of irrigation as a causal factor behind the high incidence of sharecropping tenancy is provided by a set of districtwise data on irrigated and unirrigated area cultivated by subtenants on different rates of produce rent. These data were made available in the *Punjab Land Revenue Administration Reports* till 1900-1. Statistics relating to this last year reveal that some 42 per cent of the total irrigated area of the Greater Punjab (i.e. British Punjab and the N.W.F.P.) was operated by sharecroppers but in the case of unirrigated area such land accounted for only about one-fifth (i.e. 22 per cent).[38]

TABLE 4.5: CORRELATION BETWEEN IRRIGATED AREA AND INCIDENCE OF TENANCY

Serial Nos.	*Variable Names*	*r*	*Level of Significance*
1.	per cent of irrigated area and per cent of area cultivated by subtenants paying rent in kind and cash	0.52	0.01
2.	per cent of irrigated area and per cent of area cultivated on sharecropping terms	0.49	0.01
3.	per cent of canal lands and per cent of area cultivated by subtenants paying rent in cash and kind	0.58	0.001
4.	per cent of canal lands and pper cent of area cultivated on sharecropping terms	0.52	0.01
5.	per cent of canal lands and per cent of families owning 25 acres and more	0.75	0.001
6.	per cent of families owning 25 acres and more and per cent of area cultivated by subtenants paying rent in kind and cash	0.49	0.01
7.	per cent of families owning 25 acres and more and per cent of area cultivated on sharecropping terms	0.37	0.10

Sources: Data presented in Table 4.4.

Why did the lessors, by and large, prefer produce over cash rent even though the latter mode of surplus appropriation was obviously inconvenient? The explanation lies, primarily, in two considerations. Firstly, as mentioned earlier, the Punjab *bataidars* (i.e. lessors) were paying a part of the cost of production. This meant that the *bataidars* or landowners had a stake in the productivity of land. Such a situation should, in itself, have encouraged them to prefer produce as against cash rent. But this tendency was even stronger because agricultural prices were increasing over time. Secondly, yield rates of certain crops (already higher in irrigated than in unirrigated lands) were improving as a result of the introduction of new varieties of seeds. Thus, overall, the large grantees (except perhaps those who were not resident in the villages) and the landowners preferred produce rent because it meant an automatic increase in their share along with enhanced productivity and rise in the price level. Cash rent would have served the same purpose if the rate could be adjusted annually, but obviously such an arrangement would have been cumbersome and, therefore, unworkable. There are three sets of evidence (all indirect in nature) to suggest that this explanation is a valid one. Firstly, cash rent was predominant in the dry areas of the province. For example, it was reported that "cash rates were at once most common and most stable"[39] in the south-east Punjab where crop yield was subject to wide fluctations and crop failure rate very high. Secondly, as agricultural prices drastically declined in India with the onset of the Great Depression (1929) many landlords who had earlier leased-out on produce rent, wanted to switch over to cash rent. But the subtenants refused to accept such an arrangement.[40] Thirdly, as mentioned earlier, sometimes the *Thekedars* (cash lesses) of vast areas did not themselves cutivate lands, they leased-out the "borrowed" lands to the sharecroppers.

Two other questions need to be addressed before we proceed further: (a) Why did the lessors agree to bear a part of the cost of production though, their land being much in demand, they were in a position to claim a high rate of rent? We can not argue that the *bataidars* were only conforming to an age-old custom because though under the Sikh rule and land revenue was mostly collected in kind the state could not have borne a part of the cost of production. (b) Should the sharecropping system be characterized as "semi-feudal"?

The first question does not seem to have exercized the mind of the members of the bureaucracy though they gave considerable attention to matters relating to tenancy and rent. As about the presentday commentators both Mridula Mukherjee and Neeladri Bhattacharya are silent about this problem. But the explanation does not seem to be far to seek. For though there was considerable demand for land from the *kamins* as well as the marginal and small farmers, their resources were not adequate for undertaking cultivation without the assistance of the large landowners/grantees. The latter groups were not unwilling to offer such assistance through cost-sharing arrangement because they were also keen to lease-out the whole or part of the land they possessed. As the external demand for Punjab's agricultural produce increased, rail and road communication improved and additional land became available with the expansion of irrigation network, the large landowners/grantees found it in their interest to agree to such an arrangement. This is not to suggest that the tenants had a "scarcity value",[41] i.e. the supply of land was greater than the available demand. This was not actually the case even in the canal colonies where vast (crown waste) lands were being brought under cultivation. For example, it was reported from Lyallpur districts that there was "no difficulty in obtaining tenants, in fact it is the other way round, the tenants experience difficulty in getting as much land as they want for cultivation".[42] The point is that the landlords had certainly a stake in helping the tenants bring their land under the plough.

The view that *batai* cultivation of Punjab should be characterized as "semi-feudal" has been advocated by Mridula Mukherjee. She takes this view "not simply because this form has traditionally been associated with feudalism, but because this form of surplus appropriation was here utilized to maintain semi-feudal semi-colonial conditions in agriculture".[43] By way of elaboration she adds:

> Surplus appropriation through *batai* etc. led not to accumulation and investment of capital nor to rapidly increasing productivity and shift in the basis of cultivation to a higher level, nor to a transformation of the organization of production. It led conversely to accumulations which found their way to usury, land mortgage, land purchase for further leasing out on *batai* and tenancy, leading in turn to the prepetuation of petty commodity production at more or less the existing levels of production.[44]

There is nothing new in this characterization; earlier Bhaduri took the same view with regard to tenancy in Eastern India. Indeed, a large body of scholarly opinion is in favour of such a characterization of sharecropping tenancy in underdeveloped agriculture. But, there are some commentators who reject this view and argue that actually sharecropping represents a particular form of capitalist relations.[45] Without going into the details of many issues raized in this debate we would argue that though sharecropping system has been in many cases an institution with marked pre-capitalist features, the system as it functioned in the Punjab should not be regarded as semi-feudal. Firstly, the Punjab landowners paid a part of the production cost. Whatever might have been the initial considerations for agreeing to such an arrangement, cost-sharing made a significant difference in the sense that the landowners/grantees had a stake in increasing the productivity of land and labour. Incidentally, we have direct evidence to show that they intervened in the labour process by "dictating" a particular crop-mix to their tenants and encouraging the use of new varieties of seeds. Moreover, some of them made interest-free advances to enable the tenants to bear their part of the input costs. Such practices as cost sharing, selection of crop-mix and interest-free advances are foreign to feudal or even semi-feudal landlords. It is true that instead of launching direct cultivation on a large scale or trying to improve productivity at a rate higher than what was achieved, the landowners invested in moneylending with or without land mortgage. But there was nothing peculiarly feudal or semi-feudal about such a tendency, it only shows that Punjab agriculture did not experience a capitalist transformation or a Green Revolution. Secondly, as pointed out by Mridula Mukherjee herself,[46] "debt-bondage" which has been taken by Bhaduri as a defining feature of semi-feudalism was absent in the Punjab. As a matter of fact we would argue that *batai* cultivation as it prevailed in the Punjab was an intermediary form between capitalist and (in the absence or a better phrase) non-capitalist methods of production. It was not fully non-capitalist in the sense that while in some respects the *bataidar* resembled the capitalist farmer, the sharecropper was somewhat like a wage labourer. Nor was it fully capitalist because the landowners did not accept full operational responsibility and the sharecroppers were not separated from the means of production.

4.3 Further Comments on the Absence of Capitalist Transformation

It has been briefly mentioned why, overall, we did not expect the large grantees in the canal colonies as well as the large landowners in the other irrigated (and nonirrigated) tracts of the province to prefer capitalist farming with the help of hired labour to leasing-out land to the sharecroppers. However, given the importance of the subject (absence of capitalist transformation) further comments will be in order even though it will not be possible to say anything conclusive. To begin with, by the absence of capitalist transformation we do not suggest that were was no trend in this direction in Punjab agriculture during the period under review, especially in the first half of the twentieth century. This point needs emphasising because Mridula Mukherjee does not "*find any evidence*" (italics added) of either large-scale capitalist farming through hired labour, whether engaged in by capitalist farmers or capitalist tenants (the Amerian or British model), or any growth of rich peasant farming on capitalist lines (the French model)".[47] Bhattacharya, who deals with the central and south-east Punjab, may be said to take the same view. For, firstly, he does not refer to any evidence of large-scale farming by the landlords. Secondly, though he mentions that "in Central Punjab a class of rich peasantry was slowly crystallising"[48] (a view for which he is taken to task by Mridula Mukherjee) according to his definition the rich peasants derived the larger part of their income from self-cultivation with the help of family labour,[49] not by the surplus generated by hired labour.

But actually there are clear indications that a significant part of the cultivated area of the province was operated on capitalist and semi-capitalist lines. Firstly, it appears from the available data (Table 4.3) that about one-twelfth of the total cultivated land of the province belonged to holdings which exceeded 50 acres in size. These holdings (and even many of the smaller ones) must have been operated wholly or mostly with the help of hired labour and their production was primarily meant for the market. Comparable data relating to other provinces are not always available, but it seems that the proportion of such land was one of the highest in the Punjab.[50] In the canal colonies some of these capitalist farmers belonged to the group of large grantees and auction purchasers of non-agricultural origin. This is clear from the examples of the estates of Colonel E.H. Cole and Ujjal Singh. Cole's estate consisted of 7,500 acres a thousand of which were cultivated by wage labour. The estate had six tractors, a farm machinery repair shop and a stud farm. The rest of the land was leased out on *batai* terms.[51] The estate of Ujjal Singh, situated in the Lower Bari Doab Canal had 1,700 acres worked by hired labour.[52] Incidentally, as indicated by the data in Table 4.3, it is clear that there were cultivated holdings of this size (50 acres and above) also in some of the tracts outside the canal colony districts. Thus, it is difficult to accept the view that there is "no evidence" of capitalist farmings in the province.

Significant part of land was also being cultivated on semi-capitalist lines in the sense that there was a considerable growth of rich peasant farming both in the rain-fed and irrigated tracts of the province. Incidentally though Mridula Mukherjee takes this view, she finds that whatever cut-out point is taken to identify them—30 to 40 acres in the canal colonies and 15 to 25 acres in districts like Jullunder, Ludhiana, Amritsar—a certain portion of the land should be said to have had rich peasant farming.[53] Given the wide variations in water supply, rainfall, soil conditions, crop patterns, it will be naturally difficult to decide at what point the demarcation line should actually be drawn to identify the rich peasants in different parts of the province.[54] But we can safely assume that at the provincial level most of the land in the holdings ranging from 25 to 50 acres were being cultivated by the semi-capitalist operators, i.e. those who employed family labour but derived the larger part of their income from the surplus generated by the hired labour.

Thus, the available data make it abundantly clear that in the Punjab province a significant part of the land was being operated on capitalistic and semi-capitalistic lines. Secondly, such a trend was more marked in the irrigated tracts than in the dry ones. This is clear from the fact that proportionately more land was cultivated in these two categories of holdings (25-50 acres and over 50 acres) in the districts with a higher percentage of irrigated land, especially in the canal colonies. Incidentally, some of these districts (Shahpur, Sheikhpura, Multan and Jhang) seem to have witnessed what has been called "reverse tenancy", i.e. leasing-in by rich peasants who sought to expand cultivation.

But overall, there is no denying the fact that capitalism was far from being the dominant mode of production in Punjab agriculture. At the provincial level the rice peasants failed to expand their operational holdings by leasing in additional lands from other social groups.[55] On the other hand, the large grantees and landowners mostly preferred cultivation of their lands with the help of subtenants. Therefore, the questions that need to be addressed are: why (a) the role of those who owned more than 50 acres remained, by and large, "anti-productive, parasitic and generally divorced from the risk and responsibilities of production"[56] and (b) the rich peasants did not lease-in land from above (who owned more than 50 acres) and/or those from "below" (with less than 25 acres). In other words, we need to discuss why transition to capitalism did not come about in either of the two ways as mentioned by Marx.

The debate on the question as to why and how capitalist transformation took place in the West and the absence of such a transformation in Indian agriculture has generated a large body of opinions.[57] To refer to some of those on the latter question, according to one view the large landowners' preference for leasing-out land over direct cultivation was the consequence of ideological values and social distance, e.g. considerations of caste and status prevented them from direct cultivation.[58] As Adnan has aptly pointed out, if this argument is taken to its logical conclusion, then the antecedent social considerations precluding manual or even supervisory work themselves will require explanation.[59] What is also important is that these ideological factors do not explain why the rich peasants did not expand the scale of their operations in the manner suggested earlier. According to Calvert, a keen observer of the agrarian scene of the Punjab, two considerations led the big owners to employ tenants: (a) customary rent was a rack-rent which returned to the owner more than he could obtain often paying wages of hired labour and (b) it was easier to rack-rent a tenant than to exact a fair day's work from a hired labour.[60] Basically the same view is taken by Utsa Patnaik when she attributes the historical absence of capitalist development in Indian agriculture to the fact that potential rate of return from capitalist production has been lower than the corresponding rates of mercantile profit, usurious interest and precapitalist rent.[61] Krishna Bharadwaj views the problem in the same light—prevailing poverty and land hunger in the under-developed countries provide a fertile ground for exploitation through tenancy and usury and, consequently, productive investment in land does not appear attractive.[62]

For several reasons Adnan is critical of this attempt to explain the absence of capitalist production in terms of its relative unprofitability. He seeks to explain this phenomenon by arguing that since, in the given situation, the large landowners can reproduce through pre-capitalist exploitation (especially rent) they can afford not at all to undertake production on capitalist lines or to initiate such production only to a limited extent.[63] Adnan stresses the importance of the avenues of pre-capitalist exploitation because this means that the landowners "do not go out of business" even if surplus derived from such sources of profit fail to match (a) actually realized profit rates in other sectors of the economy or (b) counterfactual rates of (capitalist) rent and profit which could be obtained, if they were to "become" capitalist.[64]

Data on the shares of the landlords and tenants in gross income, cost of production and net income in respect of some of holdings were collected under the auspices of the Punjab Board of Economic Inquiry (Table 4.6) over two decades beginning from the early 1920s. The largest of these was a government Seed Farm of 800 acres operated by 16 tenants on *batai* terms. At first sight these data would appear to contradict the view that cultivation with the help of sharecroppers was more attractive or profitable. For net income of tenants appear to have exceeded the expenditure incurred by them.

In other words, these data indicate that the additional income which would have accrued to the lessors if they had undertaken direct cultivation would have been greater than the (produce) rent paid by their tenants. But several considerations suggest that these "actually" realized returns should not be treated as an index of the potential returns from direct cultivation on capitalist lines. Firstly, in calculating the cost of production borne by the sharecroppers the Punjab Board of Economic Inquiry took into consideration only that part of human labour which had to be paid for in cash. Or, to put it differently, cost of family labour was not included. But if the landowners/grantees were going to launch direct cultivation on a large scale, given the prevailing level of technology, it would have been necessary for them to employ a large number of hired labour or more or less a permanent basis.[65] Secondly, they would have to bear the cost of maintaining bullock labour. Thirdly, as mentioned by Calvert, the productivity of hired labour would have been lower than the productivity of family labour employed by the sharecroppers.[66] Fourthly, there was no deduction for land revenue and other cesses amounting Rs 7.00 which would

TABLE 4.6: SAMPLE DATA ON THE INCOME & EXPENDITURE OF THE LANDLORDS AND TENANTS (PER ACRE)

Year	*Landlords*			*Tenants*		
	Gross Income	*Expenditure*	*Net Income*	*Gross Income*	*Expenditure*	*Net Income*
1926-7	48	8	40	48	28	20
1927-8	48	10	38	48	31	17
1928-9	51	11	39	50	27	22
1929-30	46	10	36	46	31	15
1930-1	25	9	17	26	24	2
1931-2	26	9	17	26	17	8
1932-3	34	9	25	34	20	13
1933-4	29	10	19	29	20	9

Source: *Farm Accounts in the Punjab*, 1933-4 (Lahore, 1936).

have been paid by a private landlord. Overall, "actual" income from self-management or direct assumption of full operational responsibility would not have been as attractive as these data indicate. It is worth emphasizing here that data generated by Farm Management Surveys in India suggest that small holdings as well as a number of larger ones would have incurred losses if all costs were imputed.[67] It would then appear that the larger grantees and landowners in the Punjab of our period had neither the incentive nor the compulsion to undertake capitalist production on a large scale.

To turn our attention to the rich peasants, it may be mentioned at the outset that their behaviour can not be explained in terms of their supposed outwardly attitude[68] or the tendency to cease extend production after an equilibrium had been achieved between their consumption needs and the drudgery of efforts[69] or any ideological factors. For they were already engaged in cultivation with family and hired labour and the surplus generated by the latter constituted the larger part of their income. Moreover, in some cases the rich peasants were resorting to "reverse tenancy". Their unwillingness or inability to lease-in additional land and extend the scale of operations with hired labour beyond a certain point seems to have been due to two considerations. Firstly, while it would have been necessary for them to pay a high price (in terms of rent in cash or kind) for "borrowed" land, for reasons mentioned above, labour cost would have been proportionately higher. Secondly, there was much demand for credit at high or usurious rates of interest it is possible that the prevailing interest rates were or appeared to be higher than the potential returns from the employment of hired labour on "borrowed" land. Once again, this implies that the compulsion mentioned earlier for expanding the scale of operation was absent. It may be recalled here that many rich peasants had taken to moneylending in the province as elsewhere in the subcontinent.

4.4 Level of Rent

Till 1900-1 the *Land Revenue Administrative Reports* furnished districtwise statistics on irrigated and unirrigated area held by subtenants at different rates of produce rent. These data reveal a high degree of positive correlation between (a) the percentage of irrigated area in a district on the one hand and (b) incidence of sharecropping tenancy and (c) the proportion of (tenant-cultivated) irrigated area for which half or more than half of the produce was appropriated by the landowners/grantees as rent. In other words, these statistics indicate that the districts with relatively more irrigated area did not only have a higher incidence of sharecropping tenancy, but the pitch of rent was also higher in these tracts. Incidentally, the same statistics reveal that the percentage of area for which half or more of the produce was paid as rent by the subtenants was much lower in the unirrigated tracts. It is not difficult to explain why, even though the level of rent was higher in the irrigated area, there was such a demand for such lands. For, as mentioned earlier, yield rates of irrigated land was higher and more stable (crop failure being lower). Moreover the subtenants could cultivate larger holdings in the canal colonies.

What happened to the pitch of rent in the post-1900-1 period? As mentioned earlier, the greatest merit, from the landlords' point of view, of the *batai* system was its elasticity, i.e. automatic increase in rent along with the improvement in yield rates and/or price level. Now, as the yield of certain crops improved, there was a certain shift of resources to the superior crops and agricultural prices in general showed a considerable upward trend, it is obvious that the landlords' share of the surplus (as quantum of product or product prices) increased in absolute terms. What is, however, more important to note is that their share increased also in relative sense. This happened in two ways: direct and indirect. The direct method consisted of the enhancement of the share of produce without altering other conditions of tenancy. Thus, it was reported that in Gujranwalla district the *batai* rates are "tending to move upwards, more to the advantage of the owners" and in latter years area leased at rates higher than one-third *batai* was certainly more than ever before.[70] Again, about Lahore it was said that rent rate both in cash and kind was "appreciably higher than at last settlement".[71] The landlords enhanced the *batai* rent indirectly in the sense that they ceased to pay their full share of the cost of production. For example, they shifted the owners' rate (or "water advantage rate") and land revenue to the tenants, partially or fully stopped providing a share of the seeds and tried to reduce the deduction from their share of the produce for the payment of the dues to the village menials. In the late nineteenth century the dues of the *kamins* were almost invariably paid from the common heap before the division of the produce between the landlords and the tenants.[72] Another method resorted to by the landlords to augment their share of the produce was a claim to a part of the straw according to the rate of *batai*. Around 1912 a Settlement Officer pointed out "Nowadays even landlords who are absentees or who themselves own no cattle insist on receiving their share of straw which they can usually sell on the threshing floor."[73] But in the late nineteenth century the tenants normally collected all the straw after the crops were reaped. From one village in

Lyallpur district it was reported that where the landlord was resident, he took his full share of fodder and when he was not and when he did not require fodder he "rigidly exacted the monetary equivalent of his share".[74] Again, in another village survey it was pointed out that "in this respect (taking a share of fodder crops) the landlords were far more liberal earlier".[75] It may be recalled here that fodder prices were increasing fast in the province.

One may argue that the relative increase in the landlords' share of the surplus is not quite revealing, because cash rent paid by the subtenants went up in the different parts of the province, both irrigated and rain-fed. This is abundantly clear from two sets of evidence. Firstly, during the late nineteenth and early twentieth centuries there was a remarkable increase in suits for rent enhancement[76] and "applications for enhancement were almost invariable successful".[77] Secondly, there is an overwhelming body of quantitative information on the subject available from the *Settlement Reports, Assessment Reports* and *District Gazetteers*. For example, if we compare the relevent statistics available from the last *Settlement Report* of the nineteenth and the first report of the twentieth century we find the following rates of enhancement in cash rents paid by subtenants in five districts:[78] Jullunder 100 per cent, Ambala 88 per cent, Ludhiana 106 per cent, Lahore 218 per cent, Karnal 30 per cent. But it is highly unlikely that cash rent rate increased faster than the share of the *bataidar*.

Thus, to summarize, expansion of canal-irrigation and the specific manner in which the government distributed colony lands among the different social groups of the province led to a high incidence of sharecropping tenancy in the Punjab under British rule. However, the *batai* system differed from the sharecropping tenancy in provinces like Bengal in as much as the Punjab landowners paid a considerable part of the cost of production. The question why farming on capitalist lines did not develop in the province despite the presence of large landowners/grantees and as expanding market for agricultural produce was discussed by the contemporary observers of the agrarian scene. At a general level this question has also exercized the mind of the scholars in the post-1947 period. But actually the explanation seems to lie in the willingness of the vast section of the landless people and small farmers to pay high a price for "borrowed" land rather than work as wage labour. For this meant that the large grantees or landowners had neither the pressure nor the incentive for undertaking direct cultivation.

NOTES

1. For example, H.K. Trevaskis, *The Land of the Five Rivers* (Oxford University Press, 1928), p. 349; H. Calvert, *The Wealth and Welfare of the Punjab* (Lahore, 1922), p. 99.
2. B.S. Saini, *The Social and Economic History of the Punjab,* 1901-1939 (Delhi, 1975), p. 188; S. Bhatia, *Social Change and Politics in Punjab* 1898-1910 (New Delhi, 1987), p. 89.
3. In 1895 the proprietors cultivated 53 per cent of the total cultivated area in Punjab as against 22.5 per cent in North-Western Province, 18.1 per cent Central Province and 10.3 per cent in Oudh. For details see Government of India, *Notes on Land Transfer and Agriculture Indebtedness in India* (Calcutta, 1895), p. 253.
4. The cultivating owners formed two-thirds of the agriculturists. For details see A. Roberts, *Reports*

on the Census of the Punjab, 1868 (Lahore, 1870), pp. 28-32 and D.C.J. Ibbetson, *Report on the Census of the Punjab,* 1881 (Calcutta, 1883), pp. 380-2.

5. The proportion of land cultivated by the owners themselves was actually lower than indicated in this Table. For the owners included all those who received land grants in the canal colonies and before the Act of 1912 all of them were (government) tenants.
6. The area actually cultivated on *batai* terms was larger than reported in government publications. For, though, in the case of a village it was shown that 27.5 per cent of the area was held by subtenants paying cash rent actually there was none cultivating on such terms. What really happened is that the cash lessees (called *Thekedars*) of fairly vast areas did not cultivate themselves—they employed sharecroppers. For details see B.K. Madan, *Some Aspects of the Rural Economy in the Punjab* (Lahore, 1934). Incidentally *Batai* was the most important form of land revenue assessment under the Sikhs. Indu Banga, *Agrarian System of the Sikhs: Late Eighteenth and Early Nineteenth Century* (New Delhi, 1978), p. 88.
7. According to H. Calvert, "These tenants generally take less care in preparing the land for crops, plough it less often, manure it less, and use fewer implements upon it than others. They grow less valuable crops . . . make little or no effort at improving their fields . . . keep lower type of cattle. . . ." For details see his *The Wealth and Welfare of the Punjab* (Lahore, 1922), Ch. VII. See also R. Mukerji, *Land Problem of India* (Calcutta, 1933), pp. 149-69; H.K. Trevaskis, *The Punjab of Today* (Lahore, 1931), Vol. 11, p. 11; *Report of the Royal Commission on Agriculture in India.* Vol. 8, *Evidence Taken in the Punjab* (London, 1927).
8. For such a characterization of sharecropping tenancy see Mridula Mukherjee. "Commercialization and Agrarian Change in Pre-Independence Punjab" in K.N. Raj *et al.* (eds.), *Essays on the Commercialization of Indian Agriculture* (Delhi, 1985), pp. 85-8. For eastern India see Amit Bhaduri, "A Study in Agricultural Backwardness under Semi-Feudalism", *Economic Journal,* Vol. 83.
9. Saini, pp. 188-92.
10. Farm-level data collected by the Punjab Board of Economic Inquiry give us some idea about the variations in the landlords' share in cost of production (as also in net income). From these data it appears that this share varied from 10 to 50 per cent of the total cost. For details see the publications entitled *Farm Accounts in the Punjab.*
11. In 1900-1 landlords' share in respect of 37 per cent of the irrigated area cultivated under *batai* system was half or more than half of the produce, in 5 per cent of the area it varied from 40 per cent to less than 50 per cent of the produce, in 23 per cent of the area the share ranged from 33 per cent to less than 40 per cent and in the case of 29 per cent of the area it was less than one third. Six per cent of the irrigated area was cultivated by those who paid *Zabti* rent and fixed amount of produce. For details see *Report on the Revenue Administration of the Punjab,* 1900-1. Districtwise data furnished in this report give us an idea about inter-regional variations. Some idea about the intra-regional variations may be had from the *Assessment Reports* (henceforth *AR*) of the *Phalia Tahsil, Gujrat District* (Lahore, 1915), p. 24. See also the publications of the Punjab Board of Economic Inquiry under the little *Farm Accounts in the Punjab* and the reports on village surveys.
12. Neeladri Bhattacharya, "The Logic of Tenancy Cultivation: Central and South-East Punjab, 1870-1935", *Indian Economic and Social History Review,* 20, 3 (1983), henceforth Neeladri (1983). For the variations in rent with the nature of crops see *Final Report of the Third Regular Settlement* (1905-10) *of the Rohtak District* (Lahore, 1910), p. 23, henceforth *Settlement Reports* are mentioned as *SR.*
13. *An Economic Survey of Kali Gaddi Thamman, A Village in the Lyallpur District* (Lahore, 1932), p. 112; *An Economic Survey of Durana Langana, A Village in the Multan District* (Lahore, 1938), p. 191. In some cases the subtenants had to make payments for such purposes as the marriage of a son or daughter. For details see M.L. Darling, *Wisdom and Waste in the Punjab Village* (London, 1934), pp. 132-3.
14. *An Economic Survey of Gajju Chak: A Village in the Gujranwalla District* (Lahore, 1934), p. 167.
15. Neeladri Bhattacharya, "Agricultural Labour and Production: Central and South-East Punjab; 1870-1940" in K.N. Raj *et al.* (eds.), *Essays on the Commercialization of Indian Agriculture,* henceforth Neeladri (1985).

16. We get this impression from G. Blyn, *Agricultural Trends in India, 1891-1947: Output, Availability and Productivity* (Philadelphia, 1966).
17. 58 per cent in 1911 and 59 per cent in 1921 as against all-India averages of 70.9 per cent and 69.8 per cent. For details see Brij Narain, *Indian Economic Life: Past and Present* (Lahore, 1929), p. 349.
18. Karl Marx, "Genesis of Capitalist Ground Rent", *Capital*, Vol. 3 (Moscow, 1911). Much of the discussion on the subject in the subsequent period is based on Marx's View.
19. Cited by Imran Ali, *The Punjab Under Imperialism* (New Delhi, 1989), p. 95. The entire discussion on land grants is based on Chs. 2, 3 and 4 of this work.
20. H. Calvert, *The Size and Distribution of Agricultural Holdings in the Punjab* (Lahore, 1925), p. 14.
21. *Ibid.*, pp. 3-4.
22. This latter figure is derived by Richard G. Fox from data generated by Calvert's survey. For details see his, *Lions of the Punjab: Culture in the Making* (Universtiy of California Press, 1985), p. 59, Table 6.
23. Calvert conducted two enquiries, one into the distribution of cultivated and the other into the distribution of ownership holdings in the 29 districts of the province. But as far as the percentage of land in each holding size category is concerned he furnishes data for the province as a whole, but not for each district. Therefore, Mridula generates districtwise data on the proportion of land (owned and cultivated) included in the holdings of different sizes by multiplying the number of holdings by the mean of each holding size category. For example if there are 500 holdings in the category of 10-15 acres, she multiplies 500 by 12.5 acres.
24. This point is too obvious to need explaining in details. However, to mention one indication, a menial even with a small plot of land and earning a small part of his total subsistence from it preferred calling himself a "cultivator". This tendency was noted as early as in 1881: "It is exceedingly probable that in such cases, at any rate where the second occupation is of a menial in nature, it has been dropped and only agricultural occupation returned." *Census, Punjab*, 1881, Part 1, p. 381. Cited by Neeladri (1983), footnote 35. This tendency became even stronger in the subsequent years. For details see H.C. Sharma, "Caste Mobility Among the Artisan Castes of the Punjab", *Proceedings of the Punjab History Conference*, March 1979.
25. Neeladri Bhattacharya (1985), p. 122
26. According to the *Report of the Land Revenue Committee* (Lahore, 1938), pp. 33-4 an ordinary tenant with one or two plough did not earn as much as, and often less than, a permanently hired labour. It will be naturally difficult to say whether this was indeed the case, but the point that is missed in such an assessment is that a tenant enjoyed a higher social status than an agricultural labourer.
27. According to one view 90 per cent of the Punjab tenants were from the lower castes. M.K. Juergensmeyer, "Political Hope: The Quest For Political Idenity and Strategy", Part 1 (unpublished Ph.D. thesis, University of California, Berkeley, 1974), p. 87. But it seems that this is an exaggerated view.
28. H.Calvert, *The Wealth and Welfare of the Punjab*, 2nd edn. (Lahore, 1936), p. 176, henceforth Calvert (1936).
29. For details see Imran Ali, p. 20.
30. *Ibid.*, p. 20.
31. The fact that agricultural operations in the irrigated tracts was more labour-intensive is clearly shown by the findings of the Farm Management Surveys in different parts of (post-1947) India. For a summary of these findings see, Ester Boserup, *The Conditions of Agricultural Growth* (London, 1965), p. 40.
32. From data presented in Table 4.3 it is clear that older districts like Hissar and Ferozepore had high concentration of ownership holdings over 25 acres.
33. The obvious contention is that all those who migrated to the canal colonies did not necessarily sell their land, many of them must have leased-out land to the subtenants. It is important to mention here that between 1901 and 1931 some 1.6 million persons migrated to the canal colonies from other districts: 749,920 between 1901-11, 694,509 between 1911-21 and 240,398 between 1921-31. For details see the respective *Census Reports*.
34. According to Neeladri Bhattacharya's (1983) calculations 22 per cent of the small owners of village Suner and 20 per cent in Tehong leased-out their land to subtenants.
35. Tenancy could also be due to the fact that such persons held land as mortgage but did not

themselves cultivate the mortgaged land. Incidentally mortgage had usually been regarded as an important factor behind the expansion of tenant-cultivated area.

36. Percentage of land cultivated by subtenants in two of the three districts, Jhelum and Attock, (the other one being Rawalpindi), which provided the bulk of recruits of the army was higher than the provincial acerage.
37. Neeladri Bhattacharya (1983), p. 140.
38. Proportion of irrigation area has been found to be an important variable between inter-state variations in the incidence of tenancy in India. For details see Pranab Bardhan, "Variations in the Extent and Forms of Agricultural Tenancy", *Economic and Political Weekly,* 11 and 18 September, 1976.
39. Neeladri (1939).
40. *Ibid.*
41. Saini, p. 190.
42. *An Economic Suvey of Kali Gaddi Thamman,* p. 88. We get the same opinion from *An Economic Survey of Gajju Chak,* p. 160.
43. Mridula Mukherjee, p. 85.
44. *Ibid.*, pp. 85-6.
45. Filmeno V. Aguilar (Jr), "The Phillippine Peasant as Capitalist: Beyond the Categories of Ideal-Typical Capitalism", *Journal of Peasant Studies,* 17 (1989).
46. Mridula Mukherjee, pp. 61-5.
47. *Ibid.*, p. 82.
48. Neeladri (1983), p. 137.
49. I get this impression because according to him the limit of *Khudkast* (self-cultivation) was defined by agricultural stock and quantum of family labour, supplemented but not substituted by hired labour. Cultivators with holdings larger than this given limit leased out their surplus land. Thus defined the rich peasants were also engaged in self-cultivation and therefore derived the larger part of their agricultural income from family labour. Neeladri's views are cited by Mridula Mukherjee on p. 100, footnote 83.
50. According to the *Live-Stock Census of India,* 1940 (Lahore, 1940), there were 3,239 tractors in the province.
51. *Reports of the Royal Commission on Agriculture in India,* Vol. VIII: *Evidence Taken in the Punjab,* p. 495.
52. *Ibid.*, p. 379.
53. Mridula Mukherjee, p. 79.
54. According to Naved Hamid the rich peasants were those who owned 15 to 25 acres. For details see his, "Dispossession and Differenciation of the Peasantry in the Punjab During the Colonial Rule", *Journal of Peasant Studies,* Vol. 10, No. 1 (1982).
55. Data presented in Table 4.3 indicate that the area owned and cultivated in the size group 25-50 acres was more or less the same. This means that for the province as a whole there was little transaction in the lease market between the rich peasants.
56. Kamal Siddique, *The Political Economy of Rural Poverty in Bangladesh* (Dhaka, 1982), p. 210. For qualitative evidence on the lack of initiative in agricultural progress by the landlords and large grantees in the Punjab, see M.L. Darling, *The Punjab Peasant in Prosperity and Debt,* 4th edn. (London, 1947), p. 119; *Report of the Royal Commission on Agriculture in India,* Vol. VIII: *Evidence Taken in the Punjab,* pp. 587, 618, 715, 799; *AR Lower Jhang Branch* II (1909), Para 24; *AR Rakh Branch* (1913), Para 29.
57. Ashok Rudra *et al., Studies in the Development of Capitalism in India* (Lahore, 1978).
58. Andre Beteille, *Studies in Agrarian Social Structure* (New Delhi, 1974), pp. 66, 82.
59. Shapan Adnan, "Peasant Production and Capitalist Development: A Model with Reference to Bangladesh" (Unpublished Ph.D. thesis, University of Cambridge, 1984), p. 187, henceforth, Adnan (1984).
60. Calvert (1936), pp. 72-3.
61. Utsa Patnaik, "On the Mode of Production in Indian Agriculture: A Reply", *Economic and Political Weekly,* Vol. 7, No. 4 (Sept. 1972).

62. Krishna Bharadwaj, *Production Conditions in Indian Agriculture: A Study Based on Farm Management Surveys* (Cambridge, 1974), p. 31.
63. Shapan Adnan, particularly Chs. 1 and 5. See also "Classical and Contemporary Approaches to Agrarian Capitalism", *Economic and Political Weekly*, Vol. 20, No. 30.
64. Adnan (1984), p. 46.
65. Neeladri (1985).
66. For further detials see Calvert (1936), p. 175.
67. Krishna Bharadwaj, p. 31.
68. J.R. Boeke, *Economics and Economic Policy of Dual Societies* (New York, 1953).
69. Tom Kessinger, "The Peasant Farm in North India, 1848-1968", *Explorations in Economic History*, 12 (1975).
70. *Gajju Chak*, p. 174.
71. *AR Gugera Branch* 11, 1962, Para 19. See also *AR Mailsi*, 1920, *AR Khushab*, Para 48, *SR Lahore* 1935-39, p. 5 and *AR Chunian* 1915, Para 97.
72. For details see Neeladri (1983).
73. Cited by *ibid.*
74. *AR Toba Singh, Lyallpur District*, 1937, pp. 20-1.
75. *Durrana Langana*, p. 217.
76. *Land Revenue Administration Report* (annual).
77. *Ibid.*, 1909, p. 2.
78. *SR Jullunder* 1880-6 and 1913-17; *SR Ludhiana* 1878-83 and 1908-11; *SR Ambala* 1893 and 1910; *SR Karnal* 1872-80 and 1907-9. See also *SR Hoshiarpur* 1910-14, p. 26; *SR Gurdaspur* 1972, p. 19; *SR Ferozepore* 1910-14, p. 11; *SR Madot Jagir, Ferozepore District*, 1913, p. 40; *AR of Hafizabad Tahsil, Gujranwala District*, 1925, p. 24. Indeed almost every *AR* mentions that rent was increasing.

CHAPTER 5

IRRIGATION AND LAND TRANSFER

THE QUESTION of land transfer has figured prominently in the discussion of the agrarian economic history of the subcontinent in general and Punjab in particular. Prior to annexation and for some years thereafter sales of lands in this province were rare except in the immediate neighbourhood of large towns. However, conditions changed with the introduction of settled law and order, moderation of revenue rates, individualization of property rights in land, increased monetization of economic transaction and market-oriented production in agriculture. An active land market developed and by the 1870s area transferred annually exceeded 100,000 acres (Table 5.1). As the area alienated further increased during the subsequent years it became a matter of concern and embarrassment for many officials. For it seemed to contradict the colonial government's claim that under its care the province was enjoying agricultural prosperity. Simultaneously some officials warned that there was a political danger in the situation, given that a substantial part of the area sold and mortgaged was going to the moneylenders. If no remedial steps were taken, it was argued, the antagonism of the peasant proprietors would be directed against the government.[1] Initially the government refused to take any step, arguing that the facts were insufficient to "warrant interference by legislature to restrict the transfer of land".[2] But ultimately the view that legislative protection of the peasant proprietors was a necessity prevailed and the colonial government passed the Punjab Alienation of Land Act in 1900.[3] The Act forbade (a) sale of land by the agriculturists to the members of the non-agricultural classes except under special circumstances; (b) sale of land in execution of a decree; (c) the Bengal form of mortgage with a conditional sale clause. The Act recognized three kinds of mortgage to the non-agriculturists: (i) usufructuary mortgage for a maximum of 20 years, usufruct extinguishing principal and interest; (ii) usufructuary mortgage the terms of which were subject to no statutory limit with the mortgagor reserving the right of an occupancy tenant on payment of cash rent (the rent being taken as equivalent to interest) and (iii) collateral mortgage.[4]

Total sale and mortgage declined in province with the enactment of the protective legislation (Table 5.1). For sometime proportion of land purchased and mortgaged in by the non-agriculturists also declined. But land transfer continued to draw attention because much against the prediction of many who opposed the Land Alienation Act, prices maintained upward trend, the professional moneylenders were contravening provisions of this Act to acquire land from the agriculturists and a section of the local press complained that as the small and marginal farmers were being "swallowed up" by the bigger ones land transfer was creating a problem of landlessness. The question of the contravention of the Land Alienation Act became particularly important from the 1920s and ultimately this Act was amended in 1938 declaring all *benami* transaction void. With regard to land price, it did not really appreciate as much as the officially published data indicate. For in many cases price of land was "deliberately overstated to defeat the claims of pre-emptors".[5] In some cases prices were "fictitiously enhanced" in the sense that "fancy prices" were paid for building sites in urban and rural areas.[6] But the question of price rise figured particularly prominently in the discussion on the agrarian scene both in the official and non-official circles. Firstly, appreciation in land prices appeared to many of these observers as paradoxical because it was not accompanied by proportionate improvement in the productivity of land[7] or its rental value.[8] Secondly, it was believed to have had an adverse impact on the socio-economic life of the province. As one commentator put it:

> One of the most conspicuous features in the economic history of the Punjab has been the rapid and continuous rise in the price paid for agricultural land. This rise has had a marked effect on the development of the province; it has attracted the lawyer and middlemen class in general, as well as the cultivator, to invest their money in purchase rather than improvement; it has tended to encourage gambling in the future rise, which has appeared more important than increased production, it has led to the enactment of special legislation . . . and served to encourage tenancy and hindered the tenant firm rising to the position of proprietor. To it can be traced not a little responsibility for the economic backwardness of the province for it has drawn into land investment many crores of rupees that, but for the hope of future rise, might have been forced into industrial enterprise or productive use.[9]

What led to the increase in land prices? Was the Land Alienation Act really effective in checking the transfer of land from the agriculturists to the non-agriculturists? Should we really believe that landlessness was increasing over time?[10] What was the impact of the Land Alienation Act on the supply and price[11] of agricultural credit? It is then clear that several aspects of the problem of land transfer deserve scholarly attention, and, as Kessinger's work[12] demonstrates, some of these questions can be addressed even at the level of particular villages. But we have only a limited objective in this study. For we are primarily concerned with the question whether the pattern in the districts with the larger part of the sown area served by artificial water supply, especially the colony districts, was significantly different from the rest of the province. The analysis proceeds on the basis of data for the quinquennium 1919-20–1923-4 drawn from the *Land Revenue Administration Report*. The selection of the sample is arbitrary but it should enable us to make some idea about the underlying impact of the irrigation system as it developed in the province on land transfer.

At the outset it may be recalled that the peasant grantees constituted the largest single group among the recipients of colony lands and the Colonization of Government Land Act (Punjab) of 1912 provided that these grantees would acquire proprietory rights in land ten years after the grant of occupancy rights.[13] It is then clear that during the period under review all the peasant grantees did not enjoy freedom of alienation. Secondly, details about the conditions attached to *nazrana*-paying grants are not always available.[14] But in the case of Chenab Colony it was provided

TABLE 5.1: DATA ON TOTAL SALES AND MORTGAGE, ETC.
('000 ACRES)

Years	*Area Sold*	*Price per Acre (Rs.)*	*Area Mortgaged*	*Percentage of Area*		*Agriculturists' in Alienation*		*Agriculturists' Share in Acquisition*	
				Sold	*Mortgaged*	*Sale*	*Mortgage*	*Purchase*	*Mortgage*
1874-8	102	19	227	—	—	—	—	59	40
1879-82	165	18	190	—	—	—	—	72	62
1883-7	318	18	359	—	—	—	—	68	59
1888-92	338	26	375	—	—	—	—	80	68
1893-7	370	36	474	1.3	2.5	—	—	81	71
1898-00	399	42	429	1.4	2.1	—	—	79	71
1901-5	228	47	162	0.8	0.7	71	82	71	70
1906-10	217	64	219	0.8	0.7	71	82	75	76
1911-15	237	97	238	0.8	0.8	70	84	81	76
1916-20	209	156	289	0.7	0.8	70	73	81	81
1921-5	213	238	270	0.7	0.9	73	85	79	76
1926-30	213	254	263	0.7	0.8	69	86	73	76
1931-5	203	261	251	0.7	0.9	67	84	70	64
1936-9	254	241	195	0.8	0.7	69	83	75	63

Source: *Land Revenue Administration Report of the Punjab* (annual).

that these grantees could acquire proprietory rights after a qualifying period of five years.[15] If this practice had been followed in other colonies it would follow that the problem was the same in the case of a section of the *nazrana*-paying grantees because they were still (1919-23) devoid of the power of alienation. Auction prochasers were immediately endowed with ownership rights. However to the extent that a section of the largest two groups of recipients of lands were under obligation not to alienate their land the proportion of area sold and mortgaged in the colony districts should be interpreted with some reservation.

5.1 Sale and Mortgage

Average area sold and mortgaged per transaction was larger in the canal tracts than in the province as a whole (Table 5.2). The only exception was Lahore district. This pattern is not surprising because, as mentioned in the last chapter, there was a high concentration of large holdings in the canal tracts. Incidentally, it may be noted that, overall, the proportion of area sold was also somewhat higher in these districts.

With regard to land price, one would naturally presume that provision for artificial water supply was an important factor behind its appreciation in both the well and canal irrigated tracts. Indeed land price in these districts was so much high, it would be further presumed, that the percentage of irrigated area should appear as a statistically significant variable behind inter-district variations in land prices.[16] This expectation would be based on four considerations; irrigation meant (a) higher yield per acre, (b) lower crop failure rate, (c) better scope for the production of more valuable or profitable crops and (d) in some cases, double-cropping. In other words, irrigated lands were more productive and those who operated them had greater resources. But the districtwise data fail to indicate any significant correlation between the proportion of irrigated area on the one hand and price per acre sold and mortgaged on the other. What is more, in three of the colony districts (Montgomery, Multan and Jhang) and Muzaffargarh sale value of irrigated land did not appreciate to an extent whereby price per acre in these districts would exceed the provincial average. In other words, the expectation that the availability of artificial water supply whether from wells or canals enhanced land price is fulfilled only in the case of the two old districts with the highest proportion of well lands (Amritsar and Jullunder) and three of the colony districts (Shahpur, Sheikhpura and Lahore). We arrive at this conclusion whether the basis of comparison is the price of cultivated and uncultivated land taken together or only that of cultivated land.[17]

The comparison is even more unfavourable in the case of the mortgage value of land. For whereas price (per acre) fetched by mortgage in the well-lands was much higher than the provincial average, it was only in two of the colony districts (Lahore and Lyallpur) that mortgage value exceeded the provincial average and that too only marginally. Once again, the picture remains unchanged whether we proceed on the basis of price per acre of cultivated and uncultivated taken together or the price of cultivated land.

How can we explain this phenomenon that in sharp contrast to the experience in the well-lands in some of the colony districts land prices remained below the provincial average? To put the same question differently, what caused the variations in land prices within the major irrigated tracts? At the outset it may be mentioned that Lahore district represents an exceptional case in as much as recorded prices are 'distorted' by the inclusion of abnormally high prices paid for land in the vicinity of the municipal areas. Thus, it may be suggested that actually land prices were at their highest level in the two major well-irrigated districts (Jullunder and Amritsar). The explanation is not difficult to seek. Yield rates of crops were much higher in well than in canal lands and during the period under review about half of the cultivated land received water supply from wells. Secondly, in these old districts population density was also much higher and consequently competition for land was greater. In other words, in these tracts land had a "scarcity value". Thirdly, the land market was more competitive also becasue size of land offered for sale (and mortgage) per transaction was very small. For this meant that even the peasants with relatively small means could compete for land.

The task of explaining the variations within the colony districts is more complicated. For apart from the differenes in the fertility of land, development of the means of transport, population density and nature of crops grown, several other factors need to be taken into consideration. These are: nature of water supply (seasonal or perennial), size of grants, social origin of the grantees and auction purchasers, proportion of area with settled agriculture etcetera. However, some idea can be made as to why land prices remained so low in Multan, Jhang and Montgomery districts. Firstly, it is possible that land was less fertile in these districts. Secondly, nature of land grants had an adverse impact on the land market. Therefore, a brief reference to the pattern of land grants in these tracts will be in order. Multan district was served by the Sidhnai, Lower Bari Doab Canal and canals of the Sutlej Valley Project. Sidhnai canal was a seasonal one and therefore in order to make settlement more attractive it was decided that the minimum size of grants should be 50 acres. In the Nili Bar Colony created by the canals of the Sutlej Valley Project nearly half of the lands were sold by auction and the units of sale are likely to have been quite large. The area served by the Lower Bari Doab Canal also included large holdings as many grants were made for development purposes (for example, for the production of quality seeds). Montgomery district was served by Sohag-Para, Lower Bari Doab Canal and the canals of the Sutlej Valley Project. The Sohag-Para was a seasonal canal and, once again, for this reason the government decided that the minimm size of grants should be 55 acres. For reasons described above the size of grants was also large in the areas served by the other canals. Jhang district was partly served by the Lower Jhelum Canal which created the Jhelum Colony. In this colony more than half of the available land was distributed for horse-breeding purposes and about one-tenth was given to the military personnel for purposes other than horse-breeding. These grants were also large in size.[18]

In the last chapter it has been shown that in general the colony tracts had a higher concentration of the big holdings. From what we have mentioned above it is clear that in these three districts the concentration was more pronounced.[19] It would then follow that proportionately a larger part of the cultivated land was available for letting out to the subtenants (perhaps at lower rates of cash and kind rent) and consequently

TABLE 5.2: DATA ON SALE/MORTGAGE PER TRANSACTION, PRICE, ETC.

Districts	*% of Irrigated Area*	*Average Area (Acres) per*		*Area Alienated as % of Total Cultivated Area*		*% of Cultivated Area in*		*Price per Acre (Cultivated + Uncultivated)*	
		Sale	*Mortgage*	*Sale*	*Mortgage*	*Sale*	*Mortgage*	*Sale*	*Mortgage*
1	*2*	*3*	*4*	*5*	*6*	*7*	*8*	*9*	*10*
Jullunder	51	0.8	0.9	0.5	1.0	85	96	759(893)	455(474)
Amritsar	71	2.0	1.7	0.3	1.6	76	96	703(925)	344(358)
Gujranwalla	71	5.1	4.6	0.6	1.1	65	86	214(329)	104(121)
Montgomery	92	25.5	16.3	1.2	0.4	63	68	162(257)	59(87)
Lyallpur	99	12.6	10.3	0.4	0.5	89	90	574(649)	274(304)
Multan	90	9.2	15.1	1.6	1.3	55	61	159(289)	70(115)
Jhang	86	9.2	10.7	1.4	0.8	62	61	124(200)	80(131)
Sheikhpur	85	9.8	5.1	0.7	1.0	62	72	240(387)	90(137)
Lahore	80	2.9	3.5	0.4	1.2	64	88	1,299(2030)	267(303)
Shahpur	70	4.5	3.4	0.4	0.6	55	82	272(495)	98(120)
Muzaffargarh	76	7.3	4.7	4.0	2.1	22	53	54(254)	73(138)
Punjab	46	4.0	2.6	0.7	1.2	62	76	204(329)	168(221)

Source: (a) for irrigated area *Season and Crop Reports* and (b) the rest *Land Revenue Administration Reports*

Note : Figuress in parenthesis refer to price per acre of cultivated area.

demand for land on mortgage or outright purchase was not as high as in the old districts like Amritsar and Jullunder. Land market was depressed also because area offered for sale and mortgage per transaction was large (especially in Montgomery district). For this meant that though price per acre was so low average price per sale (or mortgage) was beyond the means of the vast majority of the prospective buyers.[20]

One might ask: if the average size of sale or mortgage has to be treated as an important factor behind the variations in land prices why were prices much higher in Lyallpur? This was so because there were some compensating factors. Firstly, in this district agricultural production was one of the most market-oriented. Secondly, in Lyallpur which was fully a colony district created by the Lower Chenab Canal 80 per cent of the lands were allotted as peasant grants. This meant that while, on the one hand, land available for letting out to subtenants was limted, on the other, there was a large body of substantial farmers ready to invest in land purchase. In the circumstances even though average size of sale was quite large, demand for land and consequently its price was so much higher in this district.

5.2 Redemption and Area under Usufructuary Mortgage

Everywhere redemptions were of two kinds—real and technical. Technical redemptions were those which were recorded by the *Patwaris* in cases when the mortgagors incurred more debt on the security of the same land. For example, A mortgaged land to B for Rs 200. He wanted more money and, therefore, asked B to lend another amount of Rs 50. The *Patwari* would then record two items: (a) redemption of land for Rs 200 and (b) mortgage of the same for Rs 250. In the case of real redemption the *Patwari* would record only as under (a).[21] It is obvious that technical redemption was made possible by the increase in the mortgage value of land as mentioned earlier. Neeladri's case studies shed a flood of light on the subject.[22] A person in *Bothar* village in Rohtak district mortgaged 19 acres in 1862 for Rs 240. In 1875 the same land was redeemed and again mortgagd for Rs 430. In the same village Tok Ram mortgaged 11 acres for Rs 25 in 1881. After seven years the land was redeemed and remortgaged to another person for Rs 45. The next year the land was redeemed once again and mortgaged to yet another person for Rs 98. But no distinction was made between these two categories while reporting the data (or redemption) in official publications. Therefore, we have not concerned ourselves with this aspect of the problem.

To proceed then on the basis of available statistics, we find that in the two major well-irrigated districts and the canal tracts, as indeed in the province as a whole, area redeemed during the quinquennium under review exceeded the area mortgaged (Table 5.3). In other words, more land was redeemed than mortgaged during these years when agricultural prices were about to reach their peak before the Great Depression. But there were considerable differences within these districts as regards the extent to which redeemed area exceeded the mortgaged area. In Jhang district redeemed areas exceeded mortgaged area by more than hundred per cent. On the other hand, in Lahore and Montgomery districts redeemed areas were almost of the same order. In the absence of detailed regional studies it will be difficult to say what really caused these interdistrict variations, but it can be safely asserted that the explanation can not be sought in the differences in the size of mortgage per transaction (Lahore and Muzaffargarh) or mortgage value of land (Jhang and Montgomery).

The picture that emerges from the data on accumulated area under usufructuary mortgage (as in 1923) is significantly different. For these data indicate that in the earlier years area redeemed annually was falling short of area mortgaged and this was true of the province as a whole as well as of the major irrigated districts (Table 5.3). In other words, though in the short-term (i.e. during the quinquennium 1919-23) more land was redeemed than mortgaged, the long-term trend in the past was just the opposite, with the result that every district had a certain accumulated area under usufructuary mortgage.[23] The significance of this development is not in doubt because often usufructuary mortgage meant first step towards sale and/or subtenancy and therefore loss of *izzat* for the mortgagor.[24] But within the overall pattern the pictrue was better in the colony districts in as much as the proportion of area under this type of mortgage was slightly smaller. To put it differently, though redemption did not equal fresh mortgage, redemption as a percentage of mortgaged area was somewhat higher than in the province as a whole. The only exception to this trend was Lahore. On the other hand, in Lyallpur district the proportion of mortgage with possession was only about one-fourth of the provincial average. Among the three older districcts the proportion was much higher in Amritsar and Muzaffargarh.

5.3 Relative Role of the Agriculturists and Non-agriculturists in the Land Market

As one would expect, the agriculturists as defined under the Land Alienation Act were always the more important operators (Table 5.3) in the land market in the sense that they were responsible for much the larger part of the area mortgaged in and out and area sold as well as purchased. But in terms of net gains the non-agriculturists performed better. As far as mortgage market is concerned this is quite clear from the relevant statistics which indicate that they mortgaged in more land than they mortgaged out (Table 5.3). The area that was passing under their control in this process was not substantial, but the underlying trend is clear enough. With regard to permanent alienation the published data indicate that at the provincial level the agriculturists were slightly gaining through purchase. However, there is large body of evidence to show that the professional moneylenders were resorting to *benami* transactions to acquire land from the agriculturists and thus defeating the objective of the Land Alienation Act.[25] It would therefore be reasonable to believe that the non-agriculturists gained also through purchase because they were actually buying more land (from the agriculturists) than the available statistics would have us believe. How far do the trends in the major irrigated tracts of the province conform to this pattern? Once again, our analysis proceeds on the basis of the data for the quinquennium 1919-23. We assume that the purchase of land by the non-agriculturists in contravention of the Act of 1900, if any, was minimal in the colony districts.

Let us start with the case of permanent alienation. Except in Jhang and Montgomery districts the larger part of land sold came from the agriculturists. This group also purchased the larger part of land in all the colony districts except in Montgomery. The pattern was the same also in the three older districts. As far as the question of net gains is concerned, the picture is unmistakably clear—in all the new districts agriculturists acquired additional land through private transactions. Thus, though it is highly unlikely that at the provincial level the members of the "agricultural tribes" increased their share in the total land, they certainly did so in the colony districts. This trend was particularly pronounced in Jhang district followed by Multan and Lyallpur. As far as mortgage is concerned the provincial pattern in respect of net gains by the non-agriculturists was true of all the major irrigated tracts except Jullunder, Amritsar and Lahore. In other words, it was only in these three districts that the agriculturists mortgaged in more land than they mortgaged out. However, while their gains in these districts were marginal, those by the non-agriculturists in other areas were quite substantial. Their gains were particularly pronounced in Jhang and Multan. Incidentally, it may be noted that in these two districts the non-agriculturists held the larger part of the area under usufructuary mortgage.[26]

Overall, in all the districts under consideration the agriculturists purchased more land than they sold. On the other hand, almost in all these tracts the non-agriculturists made net gains through mortgage. Needless to mention, as far as the canal colonies are concerned the purchasers and mortgagees did not, in all probability, belong to the peasant (*abadkar*) grantees, auction-purchasers or other recipients of land. For as mentioned earlier, they were already the owners' of large holdings. Even the holdings of the peasant grantees were so large that a part of these had to be operated with the help of subtenants. By and large these purchasers and mortgagees belonged to the groups who had migrated to the canal colonies as agricultural labourers and tenants. Incidentally the vast majority of them were members of the menial classes (*Kamins*) of the proprietory villages. It may also be mentioned here that a day's labour in the canal colonies was worth a good deal more than the same labour in the central Punjab. Similarly in most years a canal colony sharecropper was better off than a small proprietor of the central areas.[27]

To summarize the main points, judged by the proportion of area sold and mortgaged during the quinquennium under review the land market in the major irrigated tracts of the province seems to have been, on balance, somewhat more active than in the province as a whole. Secondly, statistics on proportion of area redeemed in these district do not reveal a uniform pattern, but it is clear that the colony districts had a smaller accumulated area under usufructuary mortgage. It would then follow that the burden of debt against usufructuary mortgage was lighter in these tracts. Darling's work contains enough data to contradict his contention that in the Punjab province prosperity and the burden of debt were positively correlated.[28] This will be yet another ground to question the validity of his argument. Thirdly, land transfer is generally believed to have been an important factor behind the high incidence of tenancy in a region. But we have argued that as far as the colony districts are concerned the more important factor was the specific manner in which the colony lands were disposed of by the government. The fact that the proportion of area sold

TABLE 5.3: RELATIVE ROLE OF THE AGRICULTURISTS AND NON-AGRICULTURISTS

Districts	*% of Area Sold*		*% of Area Mortgaged*		*% of Area Redeemed*	*% of Area Under Usufructuary Mortgage*
	By Agricul-turists	*To Agricul-turists*	*By Agricul-turists*	*To Agricul-turists*		
Jullunder	78	84	91	90	142	8.9
Amritsar	67	67	91	92	124	14.8
Gujranwalla	68	74	95	91	122	8.1
Montgomery	36	51	70	60	111	3.4
Lyallpur	69	86	94	92	111	2.3
Multan	57	70	73	29	116	8.3
Jhang	37	85	82	46	215	8.7
Sheikhpura	74	79	72	67	126	5.2
Lahore	69	78	84	87	106	11.6
Shahpur	76	80	97	79	162	4.1
Muzaffargarh	76	77	75	47	106	14.0
Punjab	78	87	86	80	113	10.0

Source: *Land Revenue Administration Report, Punjab.*
Note: Statistics on area usufructuary mortgage are those as in 1923-4.

annually and under usufructuary mortgage was so small compared to the incidence of tenancy further strengthens this contention. Fourthly, sale and mortgage values of land in the two districts with a large proportion of well-lands exceeded the provincial average. But this was not so in all the major canal-irrigated tracts. One important factor behind the contrast in the experience of the colony districts was the relative abundance of land available for cultivation by subtenants. It would, therefore, follow that though in several ways canal construction created conditions under which land prices would appreciate, by creating vast additional landed resources it also exercized a moderating influence on the market.

NOTES

1. The most well-known proponent of this view was S.S.Thorburn, *Mussalmans and Moneylenders in the Punjab* (London, 1886).
2. Cited by Inderjit Sharma, *Land Revenue Administration of the Punjab* (Delhi, 1955), p. 177.
3. For the debate that preceded this legislation see Sharma, *Land Revenue Administration*, Ch. 3, Shyamala Bhatia, *Social Change and Politics in Punjab 1898-1910* (New Delhi, 1957), pp. 227-35, Himadri Banerjee, *Agrarian Society of the Punjab 1849-1901* (New Delhi, 1982), Ch. 5; H.M. Van den Dungen, *The Punjab Tradition: Influence and Authority in Nineteenth Century India* (London, 1972), Norman G. Barrier, *Punjab Land Alienation Bill 1901* (Duke University, 1966).
4. Om Prakash, *Agrarian Legislation in the Punjab*, Vol. 1 (Lahore, 1940).
5. *AR of the Wazirabad Tahsil of Gujrat District,* 1925, p. 23.
6. *AR of the Garhshankar Tahsil of Hoshiarpur District,* 1913, pp 17-18.
7. *Report of Punjab Provincial Banking Enquiry Committee, Evidence,* Vol. 1, pp. 148, 293-4, 447.
8. Hirashima Shigemochi, *The Structure of Disparity in Underdevelped Agriculture: A Case Study of the Pakistan Punjab* (Tokyo, 1974).
9. H. Calvert, *The Wealth and Welfare of the Punjab* (Lahore, 1922), p. 97.
10. Landlessness does not seem to have been significantly increasing over time. For details see B.S. Saini, *The Social and Economic History of the Punjab* (New Delhi, 1925), pp. 183-4.
11. Even as late as the 1920s the professional moneylender "was still the most important source of credit". For details see *Report of the Punjab Provincial Banking Enquiry Committee,* 1929-30, Vol. I, p. 22. According to one view the rate of interest was falling. M.L. Darling, *The Punjab Peasant in Prosperity and Debt* (New Delhi, 1977), pp. 185-6.
12. *Vilayatpur* 1848-1968 (Berkeley, 1974).
13. In Lower Bari Doab, Upper Chenab, Upper Jhelum, Nili Bar and parts of Chenab Colonies colonization started after 1912.
14. P.W. Paustin, *Canal Irrigation in the Punjab* (New York, 1930), p. 70.
15. Imran Ali, *The Punjab Under Imperialism* 1885-1947 (New Delhi, 1989), p. 19.
16. There is a large body of evidence to show that many colonists were sending money to their "Home" districts and a part of this fund was being invested in the acquisition of land through mortgage and direct purchase. This process was considered to have been an important factor behind the rise in land prices in the many districts. The point to note here is that if the proportion of irrigated land in these districts was small this should have weakened the provincewise positive relationship that we are postulating between land price and irrigation.
17. We have calculated the price per acre of cultivated land by dividing price per acre of cultivated and uncultivated land taken together by the percentage of cultivated area mortgaged or sold in the same district and multiplying the result by 100.
18. For details see Imran Ali, Ch. 1.
19. The proportions of area held by the non-agriculturists were also higher. The proportions were (as in 1921-2): all-Punjab 13 per cent, Multan 28 per cent, Jhang 18 per cent, Montgomery 27 and Muzaffargarh 20 per cent. For details see *Report on the Land Revene Administration of the Punjab.*

20. It may be argued that average area offered for sale/mortgage per transaction was larger because price was lower. But a close look at the data makes it clear that the size of sale/mortgage was much larger than could be justified by the differences in the price level.
21. *Report on the Land Revenue Administration of the Punjab,* 1901, p. 4.
22. Neeladri Bhattacharya, "Lenders and Debtors: Punjab Countryside 1880–1940", *Studies in History*, Vol. 1, No. 2; "The Logic of Tenancy Cultivation: Central and South-East Punjab, 1870-1935", *Indian Economic and Social History Review,* Vol. 20, No. 2.
23. While statistics of transfer were recorded every year, area encumbered with usufructuary mortgage of a village was attested every fourth year. This means that area under usufructuary mortgage in the province as a whole actally represents the area of one quarter of villages in 1923-4, one quarter in 1922-3, one quarter in 1921-2 and the rest in 1920-1. For details See *Report on the Land Revenue Administration,* 1901, p. 12.
24. Darling, p. 8.
25. For details see *Report of the Alienation of Land Inquiry Committee, West Punjab* (Lahore, 1949): M.M. Islam, "Punjab Land Alienation Act and the Professional Moneylenders", *Modern Asian Studies* 29, 2 and *Report on the Land Revenue Administration, Punjab.*
26. In 1921-2 the share of the agriculturists and non-agriculturists (in percentage terms) were as follows: Jhang 1.8-7.0; Multan 1.4-5.9; all-Punjab 6.6-3.6. For details see *Report on the Land Revenue Administration of the Punjab.*
27. Richard G.Fox, *Lions of the Punjab* (University of California Press, 1985), p. 66.
28. For a detailed critique of Darling's thesis see M.M. Islam, "M.L. Darling and the Punjab Peasant in Prosperity and Debt", *Journal of Peasant Studies,* Vol. 13, No. 1 (1985).

CHAPTER 6

FINANCIAL RESULTS OF INVESTMENT IN CANAL CONSTRUCTION

EARLIER IT WAS argued that a primary objective of the British Colonial government behind its canal construction programme in India was to expand the revenue base, i.e. to extract a "financial surplus". Indeed, while trying to emphasize the importance of this objective we suggested that the British looked upon canal-irrigation in the Punjab as a commercial venture. The present chapter deals with this aspect of the problem. The chapter is divided into three parts. In the first part we briefly discuss the formulation of government policy with regard to the pricing of canal water and other related matters. The second part is devoted to an analysis of the trends, over six decades (1887-1947), in the most important source of irrigation revenue proper—*abiana* or water rates. This is followed, in the final part, by a resume of the trends in the gross and net revenue of the irrigation works classed as "Productive" or "Major". It would have been useful to present the financial results of the "Minor" or "Unproductive" canals. But we have refrained from this part of the exercise because from time to time some canals were transferred from one category to another, thus implying that the relevant statistical data are not comparable over time. Strictly speaking, we have the same problem in connection with the data on "Major" canals, but the problem is much less serious because these (Major) canals constituted the much larger part (about 97 per cent in terms of capital outlay in 1945-6) of the total canal network.[1]

6.1 WATER PRICING POLICY

The House of Commons Select Committee (1879) appointed to "Enquire into and Report as to the Expediency of Public Works in India with Money raised on Loan both as regards Financial Results and the Prevention of Famine" had recommended "that the construction of new works from borrowed money for the future be limited to those schemes which . . . are estimated to be productive, by yielding an annual income equal to the interest on the capital expended in their construction. . . ."[2] As per this recommendation of the Select Committee investment of capital borrowed in London money market on subsequent irrigation projects came to be judged in terms of a "project feasibility test" which involved:

(*a*) considering the capital cost of any canal simply to be the amount spent in its construction, (*b*) debiting the annual revenue of the project with (i) simple interest on the capital cost of the work at the commencement of the year and (ii) working expenses incurred during the year,

(*c*) crediting the project revenue account annually with direcct and indirect receipts,

(*d*) treating the difference between (b) and (c) as an indication of whether or not the canal project was paying its way,

(*e*) a period of ten years was to be allowed after the opening of the canal before applying this test.[3]

Thus, as far as the "project feasibility test" is concerned, it anticipated the use of "full-cost" pricing in the sale of canal water. For it provided that a productive canal should yield a money rate of return which would equal the annual working expenses and debt servicing charge. But in a policy statement issued by the government (1880) it was said that "the full value of canal water should be recovered for the general tax-payer by whom the capital is provided, so far as this can be done without prejudice to the interests of the landlords or those who use the water".[4] Thus, whereas the "project feasibility test" envisaged the recovery of "full-cost", the government emphasized the need for the appropriation of the "full value" of canal water and the fixation of tariff at a rate which would be roughly as high as "what the traffic could bear". To put it in different words, the government wanted that canal revenue should exceed the "full-cost" of constructing and maintaining irrigation works. The commercial motivation was stated even more clearly in 1904 and 1922. It was observed

> The introduction into a tract of irrigation at the expenses of the state increases the produce of cultivation and the government, as representing the tax-payer, is entitled to a portion of that increase (1) as the owner of water which it has brought to the fields at great cost and then sells to the cultivators and as (2) sovereign landlord entitled to as much as an enhancement of land revenue in proportion to the enhancement in the proprietory share of the produce. The fixing of price for the water is of the nature of a commercial transaction and should be regulated mainly by the price cultivator is willing to pay.[5]

Thus, whereas the emphasis of the "project feasibility test" was on the recovery of "full-cost" of canal water, the policy statement as above stressed the need for realising its "full value" from the irrigators. In a way, however, there is no contradiction between these two as these can be viewed as stipulating, respectively, the lower and upper limits of price or the floor and ceiling of the rate to be charged for water. From what we have stated as the primary objective in this chapter it is then clear that our task will be to assess the extent to which price of water paid by irrigators was above the lower limit and the government earnings from canals exceeded the "full-cost" of delivering water on to the cultivators' fields. However, let us first discuss the forms of pricing, that is the methods devised by the government for the collection of canal revenue.

There were two kinds of irrigations receipts—direct and indirect. The direct receipts comprised of (a) *abiana* or water rates charged for maturing crops and (b) revenue derived from many subsidiary services rendered by an irrigation work which were not immediately connected with the watering of crops, such as the tolls received from navigation, plantation, water-power, fisheries and other miscellaneous items. The indirect receipts included (a) owners rate and canal advantage rate as well as (b) interest on the sale proceeds of Crown Waste Lands in areas in which canal irrigation was introduced by the government. The Crown owned huge area in the colony tracts and before the construction of the canals these areas were of little or no value. But, as mentioned earlier, when the canal network came into existence, plots of Crown Waste Lands were sold either by auction or by private treaty and these transactions fetched considerable prices. The interest which accrued on these Extraordinary Receipts were treated as a part of the income derived from irrigation projects.

Abiana or water rate (technically known as occupier's rate under the Northern India Canal and Drainage Act, 1873) was the form of pricing canal water which prevailed in the province before the advent of the British. It was also the system followed in the United Provinces.[6] Water rates were recoverable only from the user of water for agricultural purposes or direct occupiers of land, irrespective of whether they were proprietors or tenants. The rates charged per acre varied from crop to crop. It also differed according to the form of irrigation, flow and lift. One factor determining the pitch of the rate was the amount of water ordinarily required for ripening the particular crops. The other factor was the additional value of the yield (per acre) of the crops which accrued to the farmers after deducting the cost of production. Thus, so far as its agricultural use is concerned, canal water was priced not according to volume supplied, but the extent of area irrigated. Volume was taken into consideration only to the extent that the farmer had to pay higher rate for crops (for example, sugarcane) which needed more waterings. Much lower rates were prescribed for different crops when lift irrigation was practised. However, for the province as a whole lift irrigation served only 3 to 4 per cent of the total irrigated area.

As the irrigation network expanded it soon became increasingly clear that such a system of pricing canal water was wasteful because it encouraged the irrigators to apply more water than was necessary for maturing crops. Consequently the irrigation officials of the Punjab, like their counterparts in the United Province,[7] felt the need for a change in the form of pricing, i.e. a switchover to volumetric assessment. The Indian Irrigation Commission (1903) which examined the question in considerable details took the same view. But it anticipated a number of problems in the way of volumetric charging: (1) The irrigators would be averse to this method because they would not know before irrigating a crop as how much exactly they would have to subsequently pay. (2) The cost of measuring water on individual fields would be prohibitive as this would require fixing of as many meters as there were fields taking water. It would be extremely difficult to minimize the number of meters because the irrigators would refuse to cooperate to take water on a collective basis with appropriate distribution of water among themselves remaining their responsibility. (3) Irrigation requirements varied between years of high and low rainfall. Therefore, if volumetric assessment was introduced the cultivators might be required to pay in a dry year as much as the yield was worth. But the payment would be nominal in a year of favourable rainfall even though the small quantity of water was no less essential for them than in a dry year. (4) Once water was brought to a particular water course from a long distance this water must be paid for because this could not be sent back. (5) There would be problem in fixing a scale of volumetric charges which "given all details of supply, the rainfall, the irrigated area" was such that it entailed no loss of revenue and at the same time was acceptable to all cultivators. In view of these problems of Commission ruled out the possibility of introducing a method of assessment with the help of meters and instead it favoured what is called a module system. Under this system long-term contracts were to be made with groups of cultivators who would be ensured a fixed quantum of water through a sluice at prescribed intervals. The charges would remain fixed in accordance, with the terms contracted irrespective of the area actually irrigated. The Commission recommended a gradual changeover to this system in the sense that in the first phase modules were to be used merely to control supplies, to ensure that the cultivators received steady and not abruptly varying supplies of water. Assessment on the basis of quantity contracted for was to be introduced later.[8]

Throughout the first half of the current century the Punjab government examined several schemes and proposals to introduce volumetric assessment. Incidentally, many observers now felt that the need for a change was even more urgent because the existing system of water pricing did not only mean loss of revenue to the government, it also led to waterlogging problem with its adverse impact on agricultural production and productivity. A viable method of water measurement depended on the design of an accurate module and several attempts were made to evolve a satisfactory gauge. However, as a result of the rise in steel and cement prices cost involved in fitting module throughout the canal tracts appeared to be prohibitive. Purely masonry outlets did not prove better either, for being too feeble these required constant attention and much expenses. What is more, there was a tendency among the irrigators to make illegal openings to bypass the measuring device. An important contributory factor to this situation was corruption among the lower officials. In 1920 the government took an initiative in forming "cooperative irrigation societies" in a few villages in Chenab and Lower Bari Doab Canals. The idea was that these committees would control the distribution of water and the assessing officials (*patwaris* and *zilladars*) would be withdrawn. But harmony among the irrigators which was essential for the successful working of such an arrangement was not forthcoming, and consequently the project was abandoned next year. A new method called the "contract system" was tried in 1930. Under this arrangement water was to be supplied at fixed rates for each harvest from a modular outlet designed to discharge authorized amount calculated on the area dependent on the outlet. A "contract rate" was worked out on the basis of the average revenue rates and water discharges of an irrigation division, which consisted of several villages. By 1936 there were 50 outlets taking water on contract basis and these served mostly the land of large owners in the Lower Bari Doab Canal. But it was soon found that the charges under this system were lower than the prevailing occupiers' rates, thus entailing financial loss to the government. Thus, overall, little progress was made in the introduction of volumetric assessment of canal water.[9]

At one stage the Punjab provincial government opined that "occupiers rates, unlike land revenue, have never had any scientific basis of assesssment. They have been determined by rule of thumb, by what seemed to men of common-sense possible and expedient to take. In the beginning the rates are pitched low with the idea of encouraging the use of water in the infancy of canal irrigation."[10] The government took this view when, following the onset of the Great Depression (1929), agricultural prices in the province registered a marked decline and consequently there was widespread demand for the reduction of water rates. We propose to review the government position on this demand at a later stage, but some comments will be in order here on these two points: (a) the system of assessment lacked scientific basis and (b) initially promotional (water) rates were introduced.

With regard to the basis of assessment, it is true that the introduction of water rates in a certain irrigated area was preceded by considerable discussion at different levels of the official circle—from the engineers of the Irrigation Department to the Governor-General. But the discussion was not based on reliable data on the relevant variables: yields rates, cost of production and water requirements of the different crops in the various canal tracts. In other words, water rates were worked out without giving adequate attention to the relative suitability of the canal areas for the cultivation of the different crops. This explains why there was hardly any systematic relationship between the proportion of area sown with a crop in these tracts and the pitch of its water rates. This problem drew the attention of Colonel J.W. Ottley, Inspector General of Irrigation, as early as in 1896. For example, he found that though water rate for sugarcane was the same (7.50 per acre) in all the canal areas this crop accounted for 15 per cent of the crop acreage in the Western Jamuna and 3 to 4 per cent in the Bari Doab. Again, while wheat was charged at the same rate in the different tracts, this crop occupied 25 per cent of the area in the Western Jamuna Canal, 40 per cent in Bari Doab and Sirhind canals and as much as 65 per cent in the Swat River Canal. However, instead of emphasizing the need for upward or downward revision of water rates in conformity with the relative profitability of a crop in the different canal areas he commended that the rates be made uniform and enhanced. He said,

> The more one studies these figures, the more one is impressed with the fact that there is absolutely no neccessity for keeping up so many classes, or for having such considerable variations in the rates for similar crops on different canals. My conviction is that the rates could easily be safely simplified and could on the whole be considerably raized.[11]

No attempt was made in the subsequent years to rationalize the water tariff in the light of the variations in the yield rates and production cost of the major crops in the various canal tracts. Instead, in 1924-5 Colonel Ottley's proposals were implemented with the introduction of virtually identical (and much enhanced) water rates. Without a careful look at the relevant data relating to the major canal tracts it will be naturally difficult to suggest whether as a result of this step the disparity in the burden of water rates (per acre) increased or decreased. But surely water prices still remained more or less "unaccommodating in the sense that they took insufficient account of variaton in power of production from differences in soil of fields etc. thereby causing either loss of potential revenue on good lands or overassessment of poor lands".[12] It is also plausible that there was also inter-crop disparity in the sense that water rates expressed as percentages of their gross value differed considerably. As mentioned later, there were other defects in the water-pricing system. But given a situation in which canal water was over-priced in some areas or in the case of certain crops and in others it was relatively too cheap one would find it difficult to reject the contention that *abiana* was not assessed as carefully as land revenue was.

The contention that the pitch of water rate was deliberately kept low also seems to be valid. However, we notice a certain difference in the approach of the officials under the central and the provincial government. The latter normally favoured low water rates and the considerations were both economic and political. With regard to the former the objective was to promote the spread of canal irrigation and thereby to ensure the flow of irrigation revenue at a certain level. Thus, while taking part in the discussion on the fixation of *abiana* in the Sirhind canal tracts, James Wilson, the Settlement Officer of Sirsa, suggested that initially the pitch of water rate should be kept low in order to develop irrigation rapidly and should be increased gradually in the subseqent years.[13] Again, when water rates to be introduced in the tracts served by the Lower Sohag and Para Canals were under discussion, the Lt Governor of the province pointed out that he shared the "Financial Commissioner's anxiety to give the new settlers as easy terms as possible".[14] The political considerations were explicitly stated in the case of the Bari Doab Canal. The Governor of the province pointed out that Gurdaspur, Amritsar and Lahore districts served by this canal were of great importance from a political point of view and therefore there were "strong (political) reasons for moderation in the total demand in this case and caution in fixing the pitch of the new occupier's rate".[15] Sometimes dissenting voice was heard. For example, while commenting on the rates proposed for Sirhind Canal area, the Lt Governor argued that there would be difficulties in enhancing rates once introduced, but no problem in lowering it if by experience the rate was found to "discourage particular crops or the use of the water supply".[16] In this connection he referred to the experiences of the north-western provinces and suggested that water rates should not be as low as proposed. However, by and large, officials under the provincial government tended to argue in favour of keeping water rates low. The approach of the central government differed in the sense that (though it approved the rates proposed by the provincial government), it emphasized the need for periodical examination of the adequacy of the rates prevailing in different canal tracts.[17] In some cases the central government explicitly stated that "owing to the lowness of the occupier's rates the government has not secured its share of the enhanced rental of land".[18]

In earlier years the occupiers were entitled to remission for failed crops or *karaba* in all the colony tracts. The Colonies Committee recommended that *karaba* should be abolished in areas where water supply was fairly assured, but retained in tracts where the supply was defective or precarious. The recommendations were reviewed in 1910 and accepted by the Punjab government the same year. Consequently, henceforth remission came to be granted only on a selective basis.[19] The Irrigation Department framed elaborate rules in this connection. Sown area for which remission was granted on this count accounted for about 5 per cent of the total canal irrigated area both in 1944-5 and 1945-6.

Owner's and "canal advantage" rates were basically the same thing in the sense that these were the shares due from the landowners on account of the enhanced value of land due to irrigation works. The contention behind imposing this levy was that the policy of setting low water rates led to an increase in the rent of irrigated lands, and therefore as the supreme landlords the government had a right to a share of the enhanced rent. "Canal advantage" rate first appeared in the irrigation accounts in 1869-70. Earlier the enhanced land revenue claimable on account of the introduction of canal-irrigation was determined by simply rating land as irrigated, that is, no attempt was made to distinguish the portion of assessment due to irrigation. Prinsep, the Settlement Commissioner, initiated the plan of dividing the assessment into two parts—the first representing revenue claimable from land in its unirrigated aspect and the second arising from the landowner's increased profit due to irrigation. Thus, under his scheme irrigated lands had to pay three rates of assessment, viz. (i) the regular land revenue (ii) canal advantage revenue and (iii) canal water rate. Prinsep's scheme was opposed by the Inspector General of Irrigation Col. Stratchey. The latter expressed his preference for an owner's rate rather than a fixed rate to be added to the dry rates when the land became irrigated. The difference between the two rates lay in the fact that the fixed revenue assessment proposed by Prinsep could only be imposed at the time of a Settlement of a district, while the owner's rate could be imposed any time. Secondly, the land revenue assessment was fixed for the term of the Settlement while the owner's rate could be revized from time to time.[20] The owner's rate was ultimately imposed under the Canal Act of 1873. In most cases owner's rate was limited to half of the difference between the rent paid before and after the introduction of canal-irrigation. In 1894 the government decided that no owner's rate would be levied in the case of canals to be constructed in the future, and owner's rate already assessed would be amalgamated with land tax at the time of revised settlement and reported under the heading "portion of land revenue due to irrigation".[21]

There were two systems of assessing water advantage rate on land watered by perennial canals. The first was the fixed canal advantage rate on area commanded (*nahri*). This rate was imposed in the tracts commanded by the older canals—Western Jamuna, Sirhind and Upper Bari Doab—and based on the cash rent payable on such lands. The difference between that rate and the corresponding rate for unirrigated land was called *nahri parta*.[22] The second was the purely fluctuating assessment on the area sown in the newer canals (Lower Chenab and Lower Jhelum canals and the canals of the Tripple Project). The fluctuating system was considered more suitable for two reasons: (a) the assessable value of land without irrigation was little or nothing and (b) in the case of these canals it was necessary for the engineers to have a free hand in varying the supply of water. It was hoped that in time when conditions of irrigation had become established fixed assessement system in force in the older colonies would be extended to newer ones.[23] Beginning from the revenue settlement of the Chenab colony the government did make some attempts to convert fluctuating assessment to fixed demand. For the government realized that it was losing substantial amount of revenue. Secondly, the fluctuating assessment was also highly inconvenient for the government as it required the surveillance of every harvest. As the Financial

Commissioner pointed out, every *patwari* had to work out several rates on each holding and in some cases, with complicated fractions, and the "result is a tangle of figures which the cultivators can not possibly understand and the superior officers find difficult to check".[24] But all attempts to introduce fixed assessment failed. This was primarily due to the opposition of the colonists. They opposed partly because the fluctuating assessment provided opportunities for lowering the revenue burden through bribery and corruption and partly out of fear that once fixed assessment was introduced irrigation officials would cease to have incentive to ensure water supply on a regular basis. As aptly pointed out by Imran Ali, the revenue system could have been reformed at a certain political cost. But the unwillingness to bear this cost led the government to continue with a "procedure that caused it financial loss, required a great deal of administrative effort, and encouraged corruption in the subordinate bureaucracy".[25]

At one stage objection was raized against the practice of treating canal advantage rate as a part of canal receipts. It was argued that this rate or "portion of land revenue due to irrigation" was

> undoubtedly an arbitrary deduction from land revenue, the effect of which is to introduce an unnatural element in the accounts. The Land Revenue is shown at a smaller figure and the irrigation receipts are swelled. On the one hand, government can say that the increase in Land Revenue, that is the burden on the agriculturist is smaller than it really is; on the other hand they can show that their undertakings in the Irrigation Department yield larger receipts and are hence very successful.[26]

Paustian observed that it was difficult to discover the basis of this third rate (water advantage revenue) since it would appear to fall naturally under one of the two assessments against land, namely the water rate proper or the land revenue.[27] But as mentioned earlier, the government clearly spelt out the basis of this levy: since canals constructed with the help of borrowed funds enhanced the value of land the government was entitled to claim (alongwith the price of water) a part of the rental income of the landowners. It would be legitimate to raise objection against the specific manner in which this rate was assessed. For, as pointed out by Ian Stone, owner's rate or canal advantage rate "was strictly a part of rent; yet it varied according to water rate, which was an element of production cost".[28] Moreover, it is true that in many cases the landowners shifted the responsibility of paying owner's or canal advantage rate to their tenants. But there is no doubt that water advantage charges should be treated as a part of total canal receipts.

6.2 Trends in Water Rates

Data on water rates or *abiana* charged for different individual crops or groups of crops in each of the canal areas or zones of the same tract, and their revision from time to time are available from the *Administration Report of the Public Works Department, Irrigation Branch of Punjab*.[29] But we propose to analyse the trends in the light of variations of rates charged for four crops in some of the largest and oldest canal areas at several points of time over the entire period. The importance of three of these crops (wheat, cotton and gram) derives from their weightage in the irrigated area.

Sugarcane, the other crop selected for the present purpose, was the most valuable crop because it yielded the highest gross return per acre. Separate rates were prescribed for flow and lift irrigation, but we have concentrated on the former because almost the entire (97 per cent) irrigation of the province was of this type. Data on the estimated gross return per annum of the individual crops grown in the canal tracts are also available from the report mentioned above, but we have no cropwise data on the amount realized as *abiana*. This means that we can try to make some idea about the burden of water rates charged for all the crops taken together, not each crop selected in this study.

To turn to the trends, the price charged for watering the four crops increased from the beginning of the period under review and reached the highest point[30] in 1924-5. But the rate of increase in the pre-depresssion years differed a great deal. Thus, whereas *abiana* for sugarcane increased most, the rate of increase was the lowest in the case of gram (Tables 6.1 and 6.2). With regard to the two other crops, the rate of increase was higher in cotton than in wheat. In other words, as far as the rate of increase in *abiana* till this time (1924-5) is concerned these crops stood in the following order—sugarcane, cotton, wheat and gram. Incidentally, it may be mentioned that gram was really a secondary crop in as much as part of it was grown in area cropped twice in a year. Moreover, gram was sometimes mixed with other crops. As about cotton, it may be recalled that the canal tracts were particularly suitable for the new (American) varieties and the expansion rate in areas under these varieties was accelerating from the early 1920s. The other feature that deserves emphasis in this connection is that though prices of these four (as well as other) crops were also improving during these years, especially from the start of the twentieth century, the upward trend did not match the increase in *abiana* or occupier's rates. In other words, water rates of these four crops increased faster than their prices.

The disparity between the two trends became even more pronounced when agricultural prices in the Punjab registered a sharp decline with the onset of the Great Depression. There was widespread demand in the province for the reduction in *abiana*. The provincial government responded to this situation by appointing a committee to examine the problem and make suitable recommendations. The deliberations of the committee, as rightly pointed out by one commentator,[31] were virtually a debate between the official and non-official members on whether or not a reduction in rates was justified and, if so, what the quantum of reduction would be. The non-official members who constituted the majority in the committee recommended a total relief of Rs 10,000,000 through reduction in rates. But the Governor-in-Council granted a relief of Rs 3,600,000. The reductions were concentrated on few important crops because it was hoped that this would give the greatest relief. As per this decision rates for rice, sugarcane, cotton and wheat (on perennial channels) were reduced by one rupee in each case (Tables 6.1 and 6.2). The rates for wheat on *kharif* and non-perennial channels were reduced by 8 and 4 annas respectively. Maize rate was reduced by 12 annas. On the other hand, fodder rate was increased by one rupee per acre. Rates of other crops remained unchanged. The revized rates were made effective from the *rabi* season of 1933-4. However, as mentioned earlier, though it was now more expensive to irrigate land sown with

TABLE 6.1: WATER RATES FOR WHEAT AND COTTON
(ANNAS PER ACRE)

Canals	*Wheat*						*Cotton*					
	1890-1	*1910-11*	*1916-17*	*1924-5*	*1933-4*	*1945-6*	*1890-1*	*1910-11*	*1916-17*	*1924-5*	*1933-4*	*1945-6*
W. Jamuna	36	64	80	84	68	68	36	64	64	88	72	72
Upper Bari	40	61	61	85	69	69	40	61	61	89	73	73
Sirhand	16	60	80	84	68	68	60	60	60	88	72	72
Sohag-Para	36	16	16	20	—	—	28	36	36	44	—	—
Sidhnai	—	53	53	—	—		40	53	53	—	—	—
Lower Chenab	—	68	68	84	68	68	—	68	68	100	84	84
Lower Jhelum	—	60	60	84	68	68	—	60	60	100	84	84
Upper Chenab	—	80	84	68	68	—	—	64	100	84	84	84
Upper Jhelum	—	—	80	84	68	68	—	—	64	100	84	84
Lower Bari	—	—	80	84	68	68	—	—	64	100	84	84
Price (annas per maund)	—	54	53	70	33	150	—	96	93	197	83	174

Source: *Administration Report of the Public Works Department, Irrigation Branch, Pnnjab,* Price Statistics are drawn from *Season and Crop Reports, Punjab.*
Note: Price figures represent the average of three years ending 1906-10, 1923-4, 1932-3 and 1944-5.

TABLE 6.2: WATER RATES FOR SUGARCANE AND GRAM
(ANNAS PER ACRE)

Canals	*Sugarcane*						*Gram*					
	1890-1	*1910-11*	*1916-17*	*1924-5*	*1933-4*	*1945-6*	*1890-1*	*1910-11*	*1916-17*	*1924-5*	*1933-4*	*1945-6*
W.Jumna	80	144	160	192	176	176	36	40	40	52	52	52
Upper Bari	96	113	144	194	178	178	24	24	24	53	52	52
Sirhind	160	160	160	196	176	176	—	—	48	52	36	36
Sohag-Para	48	36	36	56	—	—	16	16	16	20	—	—
Sidhnai	40	53	53	53	—	—	36	34	34	34	—	—
Lower Chenab	—	128	128	192	176	176	—	48	48	52	52	52
Lower Jhelum	—	160	160	192	176	176	—	—	—	52	52	52
Upper Chenab	—	—	144	192	176	176	—	—	36	52	52	52
Upper Jhelum	—	—	144	192	176	176	—	—	—	52	52	52
Lower Bari	—	—	144	192	176	176	—	—	—	52	52	52
Price (annas per maund)	—	75	76	110	53	186	—	45	46	56	32	125

Source: *Administration Report of the Public Works Department, Irrigation Branch, Punjab* (annual). Price Statistics are drawn from *Season and Crop Reports, Punjab*.

these (and other crops) canal-irrigation did not decline in the province. As a matter of fact canal lands continued to expand, though at a reduced scale.

The balance was tipped in favour of the landowners in the closing years of the period under review. Agricultural prices began to improve in the late 1930s and soon the rate of improvement accelerated to such an extent that within a few years price of wheat, gram and sugarcane far exceeded the level of 1924-5. The improvement was particularly pronounced in the case of the former two crops (wheat and gram). Cotton prices also improved, though not to the same extent. But although prices registered a marked rise, water rates announced by the government in 1945-6 were virtually those announced earlier in 1933-4. Thus, whereas a sharp decline in the price level was accompanied by only a nominal reduction in *abiana* in the Depression years, a pronounced improvement in the former (price of crops) in the early 1940s witnessed no change in the latter (occupier's rates). The political scene of the time ruled out any attempt by government to claim a part of the enhanced value of agricultural produce by increasing the *abiana* rates.

The trends in the water rates of these four crops were representative of the average rate realized by the government for all the crops taken together—occupier's rate per acre continued to increase from the beginning of the period under review, reached the highest point in the years immediately preceding the Great Depression and then declined during the last decade (Table 6.3). It is true that the magnitude of change in the average rate actually realized by the government for every acre of canal irrigated land was less pronounced (Table 6.3). The discrepancy is also easily explicable. For every year the government remitted a certain part of the water rates for failed crops and other considerations. Moreover, the *abiana* charged for other crops did not vary as much as the rates for wheat, cotton, sugarcane and gram. However, the underlying trends are the same as indicated by the water rates charged for these four crops.

With regard to the burden of water rate, we have mentioned that the gross value of the different individual crops grown every year in the canal lands was estimated by the officials of the Irrigation Department. It may be argued that since these officials were keen to emphasise how irrigation network built by the government had brought prosperity to the Punjab agriculturists they had a natural tendency to overestimate the values of the crops. On the other hand, one may argue that these values are really underestimates because the yield rates, worked out by the Agricultural Department were lower than actual yields. As mentioned earlier, it was reported that the principal exporting firms got fairly good results by assuming that wheat outrun is 33 per cent greater than official estimate.[32] Again, on the basis of returns submitted by the ginning factories it was found that standard yield of cotton during 1927-8–1931-2 was underestimated by 28.5 per cent.[33] It will be naturally difficult to attempt firm generalization on questions whether the Irrigation Department over or underestimated the gross returns of the crops, if there was a margin of error what this margin was and whether the margin remained more less constant over time in the case of all the important crops. However, it is clear that if, to be on the safe side, we assume that these figures are overestimates and the range of error was 10 per cent or so, average water rate and canal advantage rate realized by the government would not

TABLE 6.3: WATER RATES AND AMOUNTS REALIZED AS WATER RATES AND WATER ADVANTAGE REVENUE AS PERCENTAGE OF GROSS VALUE OF CROPS

Years	*Water Rates Collected per Acre (Rs)*	*Years*	*Amount Collected as Water Rates as % of Total Value of Crops*	*Water Advantage Revenue as % of Total Value of Crops*
1990-1	2.5	1887-96	5.8	0.5
1909-10	3.0	1897-1905	8.2	0.8
1916-17	3.3	1906-15	8.7	2.4
1925-6	4.2	1916-25	6.4	3.1
1934-5	4.0	1926-36	10.5	5.6
1945-6	3.6	1937-46	6.3	4.2

Source: Computed on the basis of data in the *Administration Report, Public Works Department, Irrigation Branch, Punjab* (annual).

seem to have exceeded one-eighth of the gross value of crops grown in the canal tracts except during the Depression years. On the other hand, if we proceed on the basis of the estimates of the Irrigation Department, irrigation revenue proper (i.e. water rate and share of land revenue) can not be said to have accounted for even one-tenth of the gross returns. Earlier we emphasized the commercial motivation behind the canal construction programme of the British colonial government. But given the pattern that emerges here it should also be pointed out that a large part of the increased returns from crop production canal tracts was left with the landowners. This explains why canal-irrigation expanded fast and in some cases it replaced well lands. Incidentally it may be recalled here that the government allowed the recipients of land grants in the canal colonies to acquire proprietory rights over their land on the payment of only a nominal price.

6.3 Financial Results of Productive or Major Canals

Data on the capital outlay (direct the indirect) gross receipts, working expenses, net revenue, interest charges and net profit of the public canals of the province for every year from 1860 to 1947 are available from the *Administration Report of the Public Works Department, Irrigation Branch of the Punjab*. These figures are based on the main canal of each system as the unit. From these data it is possible to attempt a detailed quantitative analysis of the performance (in financial terms) of each of the canals. But we will be primarily concerned with the productive canals as a whole.

The picture that emerges on the credit side (Table 6.4) has several important features. Firstly, gross revenue earned by the canals classed as Productive or Major increased from Rs 14.6 lakhs per year during 1860-9 to Rs 800 lakhs during 1937-46. The rate of increase was quite marked during the first four decades, but later it slowed down. Secondly, throughout the entire period under review irrigation

TABLE 6.4: ASPECTS OF FINANCIAL RESULTS
(MILLION RUPEES)

Items	*1860-9*	*1887-96*	*1897-1905*	*1906-15*	*1915-26*	*1927-36*	*1937-46*
1. Capital outlay	140	700	960	1,380	2,280	3,280	3,820
2. Gross Revenue	15	61	151	273	493	622	800
3. Working Expenses	6	24	48	83	141	186	176
4. Working Expenses as % of Gross Revenue	40	39	31	30	29	30	22
5. Net Revenue	9	38	108	190	353	436	625
6. Net Revenue as % of Total Capital	6.2	5.4	11.3	13.7	15.5	13.2	16.3
7. Interest Charges	7	25	37	45	75	121	141
8. Interest Charges as % of Gross Revenue	44	41	24	17	15	19	18
9. Net Profit	4	13	72	144	277	315	485
Net Profit as % of Capital Outlay	2.2	1.9	7.5	10.5	12.2	9.7	12.7
10. Water Rates	—	51	129	198	320	399	453
Water Rates as % of Gross Revenue	—	84	85	73	65	64	57
11. Share of Land Revenue	—	4	13	59	158	217	320
12. Share of Land Revenue as % of Gross Revenue	—	6	9	21	32	35	40
13. Gross Value of Crops	—	867	1,573	2,298	5,186	4,054	8,895
14. Gross Revenue as % of Gross Value of Crops	—	7	9.6	12	9.5	15.5	9

Source: Computed on the basis of data drawn from *Administration Reports, Public Works Department, Irrigation Branch, Punjab.*
Note: Capital represents amount invested till the end of the decade.

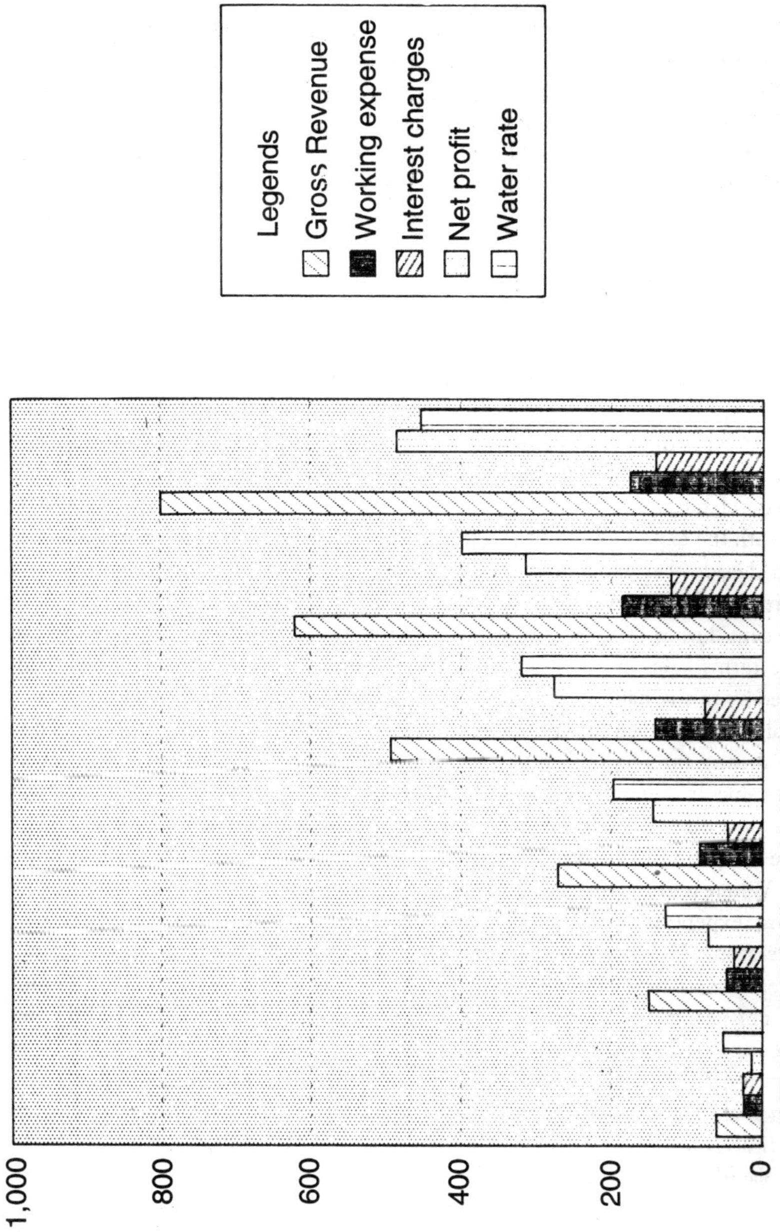

GRAPH 3: DATA ON FINANCIAL ASPECTS OF PRODUCTIVE CANALS
(LAKH RUPEES)

revenue (i.e. water rate + share of land revenue) accounted for more than nine-tenths of the total earnings of the canals. Charges for non-agricultural uses of canal water (navigation, plantation, water-power, fisheries, etc.) constituted a significant part of the total revenue only till the closing years of the nineteenth century. Thirdly, among the two sources of irrigation revenue, *abiana* or occupier's rate realized from the irrigators always remained overwhelmingly more important, but its share declined from 80 per cent during the first three decades (1887-1915) to nearly 60 per cent during the last three. On the other hand, land revenue collected as owner's rate or "water advantage rate" increased its share of the total receipts of major canals from only 6 per ent in the first decade to about 40 per cent in the last. Thus, land revenue credited to canal-irrigation increased much faster than the receipts from water rates. Fourthly, with regard to the trends during the Depression years, the rate of increase in the amount collected as *abiana* was naturally lower than in the preceding decade. For, as mentioned earlier, water rates were reduced from the *rabi* season of 1933-4 and special reduction was granted during the early years of the Depression. Moreover, canal irrigated area expanded at a slower rate during these years. What is, however, more important is that though agricultural prices ruled so low in this period, revenue from *abiana* continued to increase. The same is true with regard to "water advantage rate".

On the debit side, working expenses naturally increased althroughout the period along with the expansion of the irrigation network (Table 6.4). It was only during the last decade that expenditure under this head slightly declined. However, as a proportion of gross revenue working expenses declined from the inception, from 40 per cent during 1860-9 to 22 per cent during the last decade. In other words, in relative terms, administrative cost of delivering water onto the cultivators' fields declined by nearly half, thus indicating considerable improvement in the efficiency of canal administration. Interest rates, the other item on the debit side, also declined except during the Depression years and the rate of decline was, once again quite considerable. Incidentally, it may be recalled that interest rate on borrowed funds was 3.4 per cent till 1921 when it was raised to 4 per cent.

The Irrigation Department assessed profitability of canals in both gross and net terms. Whereas gross profit (or loss) was made up of net revenue (gross receipts—working expenses), in calculating net profit interest charges were taken into consideraton. The available data indicate that till about the last decade of the nineteenth century gross profit as a percentage of capital outlay was rather low (about 6 per cent per year). The profit rate increased during the decade 1897-1905 and further improved during the subsequent decades, reaching the highest point (16.3 per cent during 1937-46). In absolute terms accumulated gross profit stood at 1766 million as against only 14.3 million in 1860-1. During the same period total capital expenditure increased from Rs 11.3 million to Rs 410 million. In other words, accumulated gross profit in 1945-6 was more than four times than the direct and indirect capital invested in canal construction.

Net profit (gross receipts—working expenses—interest charges) as a proportion of total capital remained rather low till about the closing years of the nineteenth century. The trend in the subsequent years was the same as in gross profit, but the

TABLE 6.5 : TOTAL CAPITAL OUTLAY AND ACCUMULATED NET PROFITS (1945-6) (LAKHS OF RUPEES)

Canals	*Capital Outlay*	*As % of the the Total Capital*	*Accumulated Net Profit*	*Accumulated as Multiple of Capital*
Western Jamuna	204	5.0	944	463
Sirhind	267	6.5	1,026	384
Upper Bari	221	5.4	1,568	710
Lower Bari	240	5.9	1,818	758
Upper Chenab	433	10.6	161	37
Lower Chenab	502	12.2	5,878	1,171
Upper Jhelum	472	11.5	-348	-73
Lower Jhelum	267	6.5	1,108	415
Sutlej Project	888	21.7	285	33
Haveli Project	378	9.2	206	54
Total	4,099	—	12,881	314

Source: *Administration of the Public Works Department (Irrigation Branch)*, 1945-6, Table II-C.

magnitude of improvement was more marked than in the latter. Thus, whereas the gross profit rate increased from 6.3 per cent in the first decade to 16.3 per cent in the last, the corresponding figures for net profit were 2.2 and 12.7 (Table 6.4). The overall performance of the major canal is indicated by in figures on accumulated net profit (Table 6.4). Two points may be emphasized in this connection. Firstly, the breakeven point was reached in 1907-8 (capital outlay Rs 112 million and accumulated net profit Rs 116 million) implying that net profit from the irrigation works up to this year equalled the total capital outlay. Secondly, by 1945-6 net profit reached a point where it exceeded the total capital by more than 200 per cent (Table 6.5). In other words, every rupee invested by the colonial government in canal construction yielded, by 1945-6, more than three rupees in net terms. This was so despite the fact that gross revenue of the public canals hardly accounted for 10 per cent of the gross value of the crops produced in the irrigated area.

Among the individual irrigation works Lower Chenab Canal proved to be the most remunerative one in terms of capital/net profit ratio and its contribution (46 per cent) to the total accumulated profit. It may be recalled that the working expenses of this canal was one of the lowest in the province. The next two works in order of profitability were the Lower and Upper Bari canals. Profit rates in three other canals—Western Jamuna, Lower Jhelum and Sirhind—exceeded the provincial average. The four other important canals belonged the opposite group. In the Upper Chenab Canal the surplus was rather nominal. On the other hand, the Upper Jhelum Canal always incurred losses. These two canals were very expensive to construct, of these cut across natural lines of drainage in the *doabs*, thus requiring several training works, syphons and overhead passages. This naturally involved high interest payments. The working expenses of these canals were also proportionately higher.

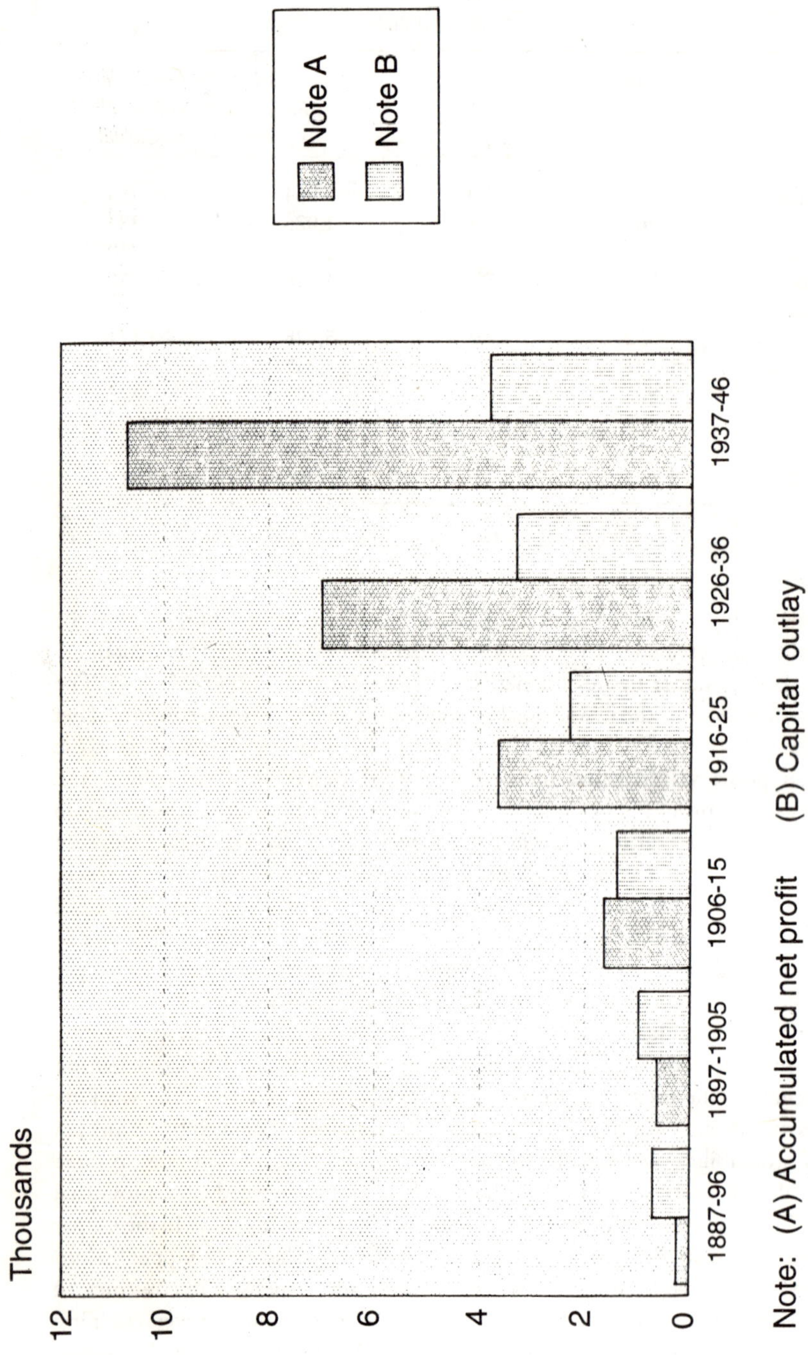

GRAPH 4: ACCUMULATED NET PROFIT AND CAPITAL OUTLAY AT THE END OF SUCCESSIVE DECADES (LAKH RUPEES)

TABLE 6.6: TOTAL REVENUE AND IRRIGATION RECEIPTS OF THE PUNJAB GOVERNMENT (MILLION RS)

Years	*Total Revenue Receipts*	*Irrigation Receipts*	*Irrigation Receipts as Percentage of of Revenue Receipts*
1921-2	86.5	33.3	38
1922-3	94.3	36.4	39
1923-4	102.2	39.2	38
1924-5	108.6	43.0	40
1925-6	113.9	45.7	40
1926-7	108.7	38.0	35
1927-8	109.0	40.5	37
1928-9	104.3	37.4	36
1929-30	105.4	39.4	37
1930-1	100.1	36.1	36
1931-2	96.1	38.1	40
1032-3	98.7	36.8	37
1933-4	103.8	44.2	43
1934-5	105.0	40.2	38
1935-6	102.4	39.9	39

Source : *Agricultural Statistics of the Punjab 1901-36* (Punjab Board of Economic Inquiry, Publication No. 52), p. 14.

The low profitability of the Sutlej Valley Project which was the biggest irrigation work in terms of capital cost can be easily explained. Firstly, this project was more expensive to construct. Secondly, it bore higher (6 per cent) interest rates. Thirdly, the working expenses were also very high. Lastly, this project incorporated several inundation canals. With regard to Haveli Canal it may be recalled that this incorporated several inundation canals and was opened in 1939. Thus, the performance of the different canals varied a great deal depending on physical and economic factors. But, to repeat, in aggregate terms the Punjab canals proved highly remunerative and these vastly added to the financial resources of the government (Table 6.6).

To summarize the main points, irrigation revenue proper, i.e. rates charged for the use of canal water for agricultural purposes, constituted the most component of the total canal receipts. Secondly, water was priced on the basis of area irrigated volume; was taken into consideration only to the extent that higher rates were prescribed for crops which required a larger number of waterings. There was a convergence of opinion that this system encouraged over-irrigation and in order to get a better return for investment in canal construction the government should introduce a system whereby water would be priced according to volume supplied. But hardly any progress was made in this direction. Thirdly, water rates in the different canal tracts did not conform to their suitability (in ecological terms) for the production of different crops. Fourthly, water rate realized per acre almost doubled between the late nineteenth century and the mid 1920s. Finally, even together with the water advantage rate *abiana* fell far short of the "full value" of canal water. In other words, larger part of the enhanced produce of land due to an assured water supply was left with the landowners. But yet by 1945-6 the government earned a total net profit of Rs 128 million from an investment of the order of Rs 410 million. Incidentally, receipts from irrigation works accounted for nearly two-fifths of the total revenue receipts of the provincial government during the period 1921-2–1930-6.

NOTES

1. There were two other categories of canals known as Protective and Provincial. Till 1900-1, i.e. the creation of North-Western Frontier Province, statistics relating to the only Protective canal of the Punjab (Swat River Canals) were furnished along with those of the Productive canals. No adjustment has been made for the separation of this canal from 1901-2.
2. Cited by B.D. Kanetkar, "The Core of Price Policy for the Sale of Canal Waters", *Artha Vijnana*, Vol. 5, No. 2 (June, 1963).
3. *Ibid.*
4. *Ibid.*
5. *Ibid.*
6. In 1921 Chief Engineer of Irrigation Works of the province argued that this system of pricing water was introduced in the early years of colonization as an incentive for breaking new ground. But I have found no evidence to support this view.
7. Ian Stone, *Canal Irrigation in British India: Perspectives on Technological Change in a Peasant Economy* (Cambridge, 1984), pp. 180-4.
8. *Report of the Indian Irrigation Commission*, 1901-3, Part 1, *General*, pp. 90-8.
9. This part of the discussion is based on Imran Ali, *The Punjab Under Imperialism* 1885-1947 (New Delhi, 1989), pp. 175-82.
10. *Report of the Abiana Committtee* (Lahore, 1934), p. 10.
11. Prog. P.W.D.C.W.I., August 1896, Nos. 19-21A. He was also of the opinion that "it is climate and locality and not the pitch of rate that really affects the cultivation of a crop".
12. Cited by Inderjit Sharma, *Land Revenue Administration in the Punjab* 1849-1901 (New Delhi, 1985), p. 55.
13. Prog. P.W.D.C.W.I., January-June, 1884. He seems to have shared the opinion of an irrigation official of the United Provinces who said, "I consider it of great importance to the Revenue to encourage in all reasonable degree, the surface spread of irrigation, for when once its advantages have been realized the cultivators will cling to them tenaciously and in course of time government will find that water has no so risen in value as fairly to autorize an increase in the rates of payment for it." Cited by Ian Stone, p. 162.
14. Prog. P.W.D.C.W.I., July 1887, Nos. 19-21A.
15. Prog. P.W.D.C.W.I., March 1891, No. 25A.
16. Prog. P.W.D.C.W.I., January to June, 1884.
17. Prog. P.W.D.C.W.I., October 1881, Nos. 911A.
18. The government took this view while commenting on the revised rates proposed for the Bari Doab Canal area. For details see Prog. P.W.D.C.W.I., March 1891, Nos. 25-30A.
19. *Punjab Colony Manual* (various editions).
20. For details see Sharma, pp. 55-61.
21. Prog. P.W.D.C.W.I., January 1894, Nos. 20-1A.
22. "By degrees the term *Nehri Patra* came to used when it was fixed a rate and if the charge happened to be fluctuating the term water advantage land revenue began to be used for that. As a rule where water advantage land revenue was imposed it was immediately paid by the occupier and where

the fixed *Nehri Patra* was taken it was generally paid by the land owner." Sharma, p. 57.

23. *Punjab Settlement Manual,* 3rd edn. (Lahore, 1915), p. 217.
24. Cited by Imran Ali, p. 169.
25. *Ibid.*, p. 174.
26. C.N. Vakil, *Financial Developments in Modern India* (Bombay, 1925), p. 245.
27. P.W. Paustian, *Canal Irrigation in the Punjab* (New York, 1930), p. 126.
28. Ian Stone, p. 167.
29. An annual publication of the Punjab government. Title varies.
30. The occupier's rates were increased to obviate the budget deficit of the past few years, the increased amount being Rs 135 lakhs or 17 per cent over the previous revenue. For details see *Abiana Committee Reports,* p. 15.
31. Nasim Ansari, *Economics of Irrigation Rates: A Study in the Punjab and Uttar Pradesh* (New Delhi, 1968), pp. 36-7.
32. *Season and Crop Reports Punjab* (1922-3).
33. *Report on the Marketing of Wheat in the Punjab* (Lahore, 1939), p. 27.

CHAPTER 7

SUMMARY AND CONCLUSIONS

ARTIFICIAL WATER supply had always played an important role in the agrarian life of the Punjab province during the ancient and medieval times. But under the British colonial rule the province was transformed into an Hydraulic Society *par excellence*. This was possible primarily because the combination of British capital and engineering skill, and local labour force brought into operation a canal network which ran into some 2,000 miles and annually irrigated some 12 million acres during the decade 1937-46. The census officials did not really much exaggerate when they observed that whereas in the past "it was only the existence of the five rivers rather than their utility that conferred the name of the province", these were now "rendered a valuable asset by the engineers' skill. . . ."[1]

During the first three decades after the annexation of the province the British confined their attention to the repair, renovation and extension of existing canals. Then beginning from the early 1880s, the government launched a programme of constructing canals with the help of funds borrowed in London money market. Railway construction accounted for the larger part of the fund raized in this manner, but Punjab received the bulk of the amount earmarked for canal construction. In all nine major irrigation works were built. Situated in the interfluves west of the Beas-Sutlej and east of the Jhelum rivers these canals were perennial, i.e. these drew off from river-spanning weirs and headworks. This made it possible to irrigate the upland plains that had hitherto remained inaccessible to the small-scale traditional irrigation methods. Incidentally through "construction of weirs of moderate height was successfully accomplished by the natives long before the English engineers came to the country to benefit by their experience and to improve on their methods of construction"[2] all the canals except one (i.e. the Hasli Canal) which the British repaired and renovated were of the inundation type.

Canal construction in the Punjab differed from similar works in other parts of the subcontinent not only because the scale was much larger, but more fundamentally in the sense that the government had to launch a colonization programme. For, the Punjab canals were built in areas which were practically deserts with little settled population. Thus, the government had to undertake a two-fold responsibility: construction of canals and an equally massive and, in a way more difficult, task of colonization in the unclaimed lands (taken over as Crown Waste) commanded by them. Colonization incorporated the different social groups in the province. Even the members of the commercial and trading classes who were otherwise barred (under the Land Alienation Act, 1901) form the purchase of land from the agricultural tribes, were allowed to acquire colony lands in the auction market. The only exception was the vast landless stratum of the rural population. Among the grantees the most important were the peasant proprietors and landlords of older districts. Thus, the colonization programme gave a powerful boost to the growth of a class of rich peasants and strengthened the position of the landlords. The position of the peasant grantees further improved when the government gave them proprietory rights and thus abandoned all its control over them in such matters as their residence in the colony, succession and alienation of land. The fact that the two rural classes grew in strength and influence had great political significance for the future during British rule and even after.

Large-scale intervention by the colonial government for the development of an irrigation infrastructure involving a colonization programme during the "high noon" of *laissez faire* principle cannot be explained in terms of government concern for the prevention of famine, reduction of population density in the older districts, creation of villages of a superior type or the "dynamism or sense of duty of the Punjab School of Administration" or the "Paternalistic ethos" of British rule. As a matter of fact canal (and railway) construction was undertaken by Government for more basic considerations: expansions of its tax base and the development of India's productive powers in the agricultural sector so that she could better fulfil her role as the supplier of agricultural raw materials, pay for her imports from Britain and earn an export surplus. Punjab received the lion's share of the investment in this field because of the convergence of several favourable factors.

However, though public-financed canal network expanded fast and soon became the more important part of the irrigation system of the province, wells continued to play a significant role in the agricultural life. For as late as the last decade of the period under review well-lands accounted for about one-fourth to the total irrigated area. This was possible because the replacement of well-lands which happened during the first three decades or so was never as pronounced as in the Uttar Pradesh. What is more, after an initial contraction well irrigated area soon began to expand in absolute terms. The most important contributory factor behind this trend was that the method of lifting water was improved in several respects: (a) while *charsa* was more common till about the close of nineteenth century, as the twentieth century progressed Persian Wheel became the more widely used water lift; (b) operational efficiency of Persian Wheel further improved with the replacement (in part or full) of wood by iron as construction materials, earthenware buckets by aluminium ones and in some cases use of electricity or oil engines by bullock power; (c) a certain number of wells was bored every year to augment water supply; and finally (d) some progress was made in the introduction of tubewell irrigation. However, except in cases when water was lifted with the help of oil engines or electricity, well-irrigation continued to suffer from the disadvantage which had earlier led to its replacement by irrigation from canals: it was more expensive.

What was the impact of the irrigation infrastructure on the agrarian economy (especially on trends in crop production) and society of the province? With regard to the first part of the question, it is clear that the role of canal-irrigation left considerable scope for improvement. For its incremental contribution to yields and total output value was below what could be otherwise expected in the given situation. At bottom the problem was due to (a) wastage of water and (b) failure of the irrigators to apply the available supply to a more efficient use. Water was wasted becasue the Irrigation Department devized a distribution system whereby it could deliver only about half of the water that entered the canal heads on to the fields of the cultivators. Secondly, the cultivators applied more waterings to a crop than it needed. Over-watering itself had an adverse impact on production (because of reduced yield rates), but the problem became more acute where this factor coupled with the seepage of water in the main canals and distributaries led to waterlogging over a vast area. The problem under (b) is underlined by the fact that though the province witnessed fast expansion of acreage under such valuable crops as cotton, sugarcane as well as those used as fodder, in the irrigated tracts there was as yet no specialization in the production of any or all these crops taken together. This is obvious because while sugarcane fetched the highest gross (as well as net) return per acre, it never accounted for more than 3 or 4 per cent of the total irrigated area. Cotton, the next best crop according to these criteria, hardly occupied one-fourth of the area as late as in the last decade of the period under review. On the other hand, with much lower gross return per acre wheat occupied a much larger area. Thus, though the Punjab cultivators positively responded to the new opportunities opened up by the government and migrated to the colonies to take to canal-irrigation and the use of well-water in the older districts was otherwise efficient, there was no large-scale transfer of resources to the production of more valuable or profitable crops.[3]

But despite these limitations on the part of the Hydraulic Regime[4] as well as the irrigators in general, provision for assured water supply made a critical difference in the crop production sector. One would arrive at this conclusion even if the available crop statistics are taken at their face value. But, as pointed out by Clive Dewey, there are grounds to believe that the Agricultural Department did not only under-report the yield rates of crops, it also underestimated the improvement that was achieved in this respect over time. In other words, the range of error in yield rates became larger through time. It is then clear that the growth rate in all crop output (especially in the irrigated tracts) was higher than the rate indicated by the officially published data. This conclusion seems to be inescapable even though admittedly several negative factors were at work along with the positive ones, i.e. use of high-yielding varieties of seeds for crops like wheat, cotton and sugarcane and the extension of irrigation. It will be naturally risky to guess what the range of error was, but one would be inclined to suggest that at the all-Punjab level crop production increased by at least one per cent per year during 1907-47. Judged by the standard of a country like Meji Japan this was not an outstanding performance. However, within the subcontinent itself the Punjab situation was certainly better, for in other provinces except Bombay-Sind crop production virtually began to stagnate from the turn of the century.

Thus, the most important impact of irrigation on the agrarian economy of the province as a whole was that it imparted a certain dynamism in the crop production sector. This was possible in two ways, directly because irrigation extended cultivation to areas which are likely to have remained uncultivated but for the construction of the canals, gave higher yields per acre than unirrigated land even when new seeds were not used and ensured greater crop security, and indirectly in the sense that it created conditions under which the new varieties of seeds could become popular with the cultivators.

The relative dynamism of Punjab agriculture was also evidenced by another significant development—commercialization of agricultural production. Larger crop yields per acre in the irrigated area and its improvement over time coupled with the fact that land/man ratio was favourable in the colonies made agricultural production in this province one of the most market-oriented in the whole of Asia. True, benefits of increased market-oriented production agriculture were unevenly distributed among the different regions and social groups. Even within the canal colonies the sharecroppers were particularly disadvantaged because they operated small holdings and the landowners appropriated a large part of the produce. But the fact that commodity production now became more widespread is not in doubt. Incidentally, it should also be pointed out that whatever might have been the pattern in other provinces, this phenomenon (of commercialization of agricultural produce in Punjab) can not be characterized as a "forced" or "artificial" one. For the large and medium farmers who marketed their produce for "gains from trade" rather than out of compulsion, contributed the much larger part of the total "marketed surplus".

With regard to agrarian structure, the most significant consequence of the irrigation system as it developed in the Punjab was a vast expansion of area cultivated on sharecropping or *batai* terms. Spread of irrigation led to this development in two ways: (a) directly because *nahri* (and *chahi*) cultivation was more labour-intensive and (b) indirectly through the specific manner in which colony lands were disposed by the government. True, a certain part of the land owned by the landlords/large grantees was operated on capitalist lines and there was considerable semi-capitalist farming by the rich peasants. But *batai* cultivation became a dominant feature of the agrarian life of the province reportedly involving a considerable loss of efficiency in production. The fact that, by and large, the landowners in the canal districts, perferred produce to cash rent even though agricultural production was now more market-oriented and otherwise the larger economy became more monetized should not appear as paradoxical at all. There was a marked upward trend in agricultural prices and yield rates of certain crops were also improving. In such a situation payment in kind by the subtenants meant an automatic increase in the landlords' income from land. This was an important consideration especially because the landowners were responsible for a part of the cost of cultivation.

Why did agriculture in the canal colonies fail to experience an overall capitalist transformation? Our basic contention is that as far as the big landowners are concerned there was little pressure or incentive to take full operational responsibility and launch large-scale farming on capitalist lines. Pressure was absent because there was a vast pool of small and marginal farmers as well landless people in the area outside the canal districts ready to pay a high price for "borrowing" land for "self-cultivation" rather than work as agricultural labourers. Incentive was inadequate in the sense that self-management of a large holding with the help of hired labour had the risk of not being profitable at all or less profitable than leasing out land on sharecropping terms and investment in moneylending with or without land mortgage. The last two considerations also explain why the rich peasant did not extend the scale of their operations with the help of hired labour beyond a certain point. Another important factor behind their unwillingness to undertake large-scale production by leasing in land from other groups was that the "price" of such land was high.

There is not much new about the way we have tried to explain the absence of full-fledged capitalist farming. Several scholars have argued more or less in the same vein. But one important "external" factor that has received inadequate attention from them is that India did not experience large-scale industrial-urban development. This is surprising because they have often argued in the light of the "classic case" of capitalist development in English agriculture. Marx's picture of the development of capitalism as the dominant mode of production included a vigorous growth of industry along with the transformation of agriculture. Actually the two sectors of the economy were symbiotically related. "The dynamic advance of industry worked as a powerful pull factor, sucking labour away from agriculture. This factor, thanks to the colonial situation, was not strong enough in India."[5] Thus, the fact that industrialization efforts in India were weak was of great significance. It created conditions under which the necessary "compulsion" and/or incentive for undertaking direct large-scale farming remained inadequate, formation of land, labour and credit market was retarded and, above all, the overall economy remained underdeveloped.

The fact that large-scale farming was not the general pattern meant limited progress in another field—use of more advanced farm implements such as reaping machines, mechanized winnowers and steam threshers. The only development worth mentioning was the use of tractors. Simpler agricultural tools and implements which were considered to have advantages over the local varieties were more widely diffused in the canal colonies. These included ploughs, harrows, hoes drills, chaff cutters and cane crushers. We can make some idea about the progress achieved in this field from the fact that in 1929 some 14,207 Indian and 7,120 imported implements were sold in the province. However, it goes without saying that the large majority of the cultivators continued to use the traditional implements. Incidentally, majority of the cultivators also failed to adopt such improved techniques of farming as use of only pure seed, conservation and experimental use of manures, sowing of crops in lines, better rotation of crops.[6]

Two other aspects of the impact of canal construction programme should also be emphasized. Firstly, the colonization programme meant a considerable demographic shift in favour of the new districts. This happened not only because of the migration of the grantees and auction purchasers, many other groups went to the colonies to take up employment in different capacities, especially as subtenants and agricultural labourers. Between 1881 and 1941 the populations of the province increased by 52 per cent. But the rate of increase was over 100 per cent in such canal districts as Jhang, Shahpur, Multan and Sheikhpura, more then 200 per cent in Montgomery and over 2,200 per cent in Lyallpur district. Secondly, the new agricultural frontier opened up by the canals led to a considerable caste mobility in the province. As vast area became available for cultivation by subtenants, mainly on *batai* terms, and demand for agricultural labour vastly increased the menial classes (*kamins*) gave up their traditional professions and migrated to the colony districts. According to the census returns of 1931 the *chamars, chuhars* and members of other artisan classes accounted for 60 per cent of the total agricultural labourers.[7] These groups also constituted the vast majority of the subtenants.[8] Moreover, many of them were now in a position to acquire land through direct purchase and mortgage. Thus, canal construction coupled with the colonization programme that accompanied it played a big role in the dissolution of the age-old *sepidari* system. There was a significant development in another respect. The villages planned and developed by the colony officers had one *dharmashala* for all, thus forcing disparate castes to come together. Thus, even though the untouchable low caste men lived on the outskirts of the villages or even in a separate area within them they were now in a position to intermingle with members of other castes in schools and places of worship. This was bound to have a positive impact on the overall caste relations.[9]

Canal construction, together with the consequential colonization programme had naturally a profound significance for the foreign rulers themselves. For it paid them rich dividends in the political, military and, most important of all, economic fields. Firstly, land grants in the colonies to the landholding classes proved politically rewarding, because the vast economic benefits they obtained in this manner provided a sound basis for their loyalty to British rule. It is worth emphasizing here that till about the end of the British rule nationalist movement in Punjab did not gather the strength which it did in other provinces. Of particular significance in this respect was the role of the landlords, Bolstered by the allotment of colony land they dominated the Unionist Party which stood for "close cooperation" with the British in all spheres of life to "subvert the hot house growth of national independence".[10] Secondly, allegiance of the landholding groups also served vital military objectives because recruitment in the British army occurred primarily from their ranks. A more direct inducement to join the army was provided by the allotment of colony lands to the military personnel. This practice became more pronounced with the outbreak of World War I when over 420,000 acres of colony lands were distributed among 6,000 commissioned and non-commissioned officers. Thus, it is not so much the martial qualities of the people, but direct and indirect inducements given by the government which made military service very attractive in the province. Incidentally at the end of the war Punjabi troops accounted for half of the total strength of the British Indian army. Needless to mention, this army was not only used to conquer and police far-flung overseas territories, but looked upon as a source of strength within India. As late as in 1938 the Governor of the Punjab asserted, ". . . in the last resort the safety of India depends not on the attitude of the Congress, but on the loyalty of the Punjab". For, as Hailey used to observe, "so long as you have a loyal and contended Punjab peasantry, you can rely on your Indian army and can face with confidence any situation that may arise in other provinces".[11]

The canals proved highly rewarding also in economic and financial terms. *Abiana* or occupier's rates constituted the bulk of the direct and indirect receipts of the irrigation works. But these rates were not assessed as carefully as the land revenue was. One consequence of this lack of adequate care was that while in some canal tracts water was overpriced, in others it was relatively cheap. Secondly, despite the widespread perception (in some cases verified by the findings of special inquiries) that the policy of pricing water according to area served led to overirrigation of crops by the cultivators with consequent loss of revenue to the government and an adverse impact on production and productivity, little progress was achieved in the way of introducing volumetric assessment. Thirdly, corruption was widespread among the lower revenue officials. Fourthly, average water rate per acre realized by the government increased over time, but the rate of increase was rather small. Moreover, the available statistics indicate that even together with "canal advantage rate" water charge did not normally exceed 10 per cent of the total gross produce per acre. In other words, the larger or a considerable part of the additional value of the produce per acre due to the application of canal water was left with the landowners. But despite all these, investment in canal construction proved highly profitable, accumulated net profit, net profit being of the order of 1,228 million rupees in 1945-7 as against a capital outlay of 410 million rupees.

The canals turned out to be so much profitable in financial terms primarily because the surface spread of irrigation was fast. This was, in turn, the outcome of four factors: moderate pitch of water rates, vast scope for the extension of cultivation, upward trend in agricultural prices and increase in agricultural population in absolute terms in the absence of industrialization efforts. With regard to the surface spread of water it may be mentioned here that every year irrigated area in the different canal tracts far exceed the limits prescribed by the Department of Irrigation. The canal engineers remained silent about this tendency even though these limits were stipulated with a view to guarding against the problem of waterlogging. Given the difficulties in introducing volumetric assessment and the political cost in enhancing water rates the government must have found it convenient to tolerate, or even to encourage, extensive rather than intensive use of canal water. This has a certain similarity with the behaviour of the irrigators who switched over, whenever possible, from well to canal water, because though the former gave higher return per acre the latter yielded a larger gross return per holding.

Irrigation works increased the financial resources of the government in another way. Plots of Crown Waste Lands were sold either by auction or private treaty and the interest which accrued on these Extraordinary Receipts was treated as part of the canal receipts. However, as in all probability these lands would not have had any sale value but for the construction of the canals the sale proceeds should also have been treated as part of the canal receipts. Needless to mention, that would have further increased the rate of net profit mentioned above. To make some idea about the magnitude of income from this source, between 1892 and 1913 more than 40,000 acres of irrigable land owned by the government passed in to private hands.[12] The government sold another 95,000 acres during the seven years from 1915 to 1921. Prices also increased. Before 1915 the highest return to the government from a single auction was Rs 266 per acre, but after 1918 the lowest average price was Rs 294 and the highest Rs 793.

Development of the irrigation infrastructure transformed Punjab agriculture into one of the most export-oriented in the whole of Asia, thus making a significant contribution to strengthen and deepen colonial relationship between England and India. This is clear from the export statistics. Between 1890-1 and 1918-19 raw cotton export increased from 0.24 million to 2.1 million maunds and the exports of cereals and pulses from 14.0 million to 40.8 million maunds. During the same period value of total exports went up from Rs 71 million to 522 million. Exports must have increased in the subsequent years both in volume and value. The contribution of the canal network in augmenting agricultural export is obvious from the vast increase in the share of the colony districts in net export—from 40 per cent during 1890-4 to about 70 per cent during 1910-14. Thus, as aptly pointed out by one commentator, "Inundated by British capital investment the canal colonies came to be the promized land of export crop production."[13] During the earlier years of the period under review the bulk of the exports was meant for overseas territories. Latter, exports to other Indian provinces increased. But Punjab continued to play an important role in India's total overseas exports because by supplying cereals and pulses (particularly wheat) to other parts of India she made it possible for them to produce exportable cash crops in a situation where total crop acreage remained more or less stagnant in the face of increased population growth.

Overall, public investment in irrigation works proved rewarding for the British colonial government: for many years it ensured the loyalty of the dominant section of the rural population, facilitated recruitment in the army, vastly added to the financial resources of the state and strengthened the colonial relationship between England and India. At the same time it brought about significant changes in the agrarian economy and society of the province. Assured water supply increased agricultural production directly through a vast expansion of cultivated area and indirectly by facilitating/promoting certain changes in crop pattern and spread of the new varieties of seeds. Increased crop output together with its commercialization in a situation of rising prices, favourable land/man ratio and water pricing policy of the government led to the rise of a large class of rich peasants and strengthened the position of landlords. Production relations did not experience an overall capitalist transformation, but there was a certain development of semi-capitalist and capitalist farming. Moreover, the sharecropping system as it developed on a large scale in the province signified some kind of "penetration" of capital in the sphere of production. There was yet another significant development in the sense that the hold of the merchant-moneylenders was relatively weak. Thus, in several important respects the experience in Punjab agriculture was different from many or most other provinces. In other words, during British rule Punjab agriculture (especially agriculture in the irrigated tract) experienced some kind of transformation and thereby reached a stage, when under more favourable conditions a Green Revolution could be possible.

At this stage a comment on Mridula Mukherjees' contention will be in order. She regrets that some historions have a tendency to emphasize "those specific features of Punjab agriculture which *at first sight* appear to deviate from the stereotypical colonial pattern" (italics added) and exclude other "features which conform more closely to the broad pattern of agrarian developments in the country".[14] Among the select aspects she lays particular emphasis on the view that (a) some kind of self-sustaining agricultural growth was occurring and (b) this growth took the form of rich peasant capitalist farming. Her contention is that there was nothing exceptional about Punjab's peformance in the crop production sector. For despite "initial advantage in terms of relatively greater availability of irrigation and improved seeds, Punjab also began to demonstrate over time an overall stagnation".[15] She also rejects the view that some kind of capitalist farming developed in the Punjab. Further, Mridula Mukherjee is of view that this tendency to highlight select aspects of Punjab agriculture has a serious implication, because it underplays the importance of the "overarching structure" of British colonial rule. If we believe that "some kinds of long-term, self-sustaining capitalist tendencies were emerging and crystallizing" how can we, she asks, "maintain that the colonial structure in India placed basic constraints on the development of capitalism in agriculture?"[16]

Three points deserve emphasis or re-emphasis in this connection. Firstly, there are valid grounds to believe that in Punjab agricultural production increased faster than the officially published data would have us believe. But even if we proceed on the basis of these data the exceptional nature of the experience of this province can not be called in question. Secondly, the available evidence make it abundantly clear that a considerable part of the land in this province was cultivated on capitalist and semi-capitalist lines and in all likelihood the proportion of such land was one of the highest in the subcontinent. Moreover, as argued earlier, the sharecropping system as it developed in the Punjab may be said to represent some kind of a "penetration" of capital in the sphere of production. Finally, there is no scope to take our emphasis on the fact that crop production significantly increased in the Punjab as an attempt to underestimate the importance of the "overarching structure" of colonial rule. With the ascendancy of industrial over merchant capital in England in the second half of the nineteenth century there was a realization that steps should be taken to improve India's productive powers in the agricultural sector so that she could better play her dependent role. At the all-India level the efforts that were put in by the foreign government to this end were extremely limited.[17] But as far as Punjab province is concerned the government came forward in a big way to construct irrigation works and develop new varieties of seeds. The result was that the agricultural sector in this province became relatively dynamic. Increased crop production coupled with its market orientation and the rise of a class of landlords and rich peasants which was thus possible, in turn, served the interests colonial state in various ways: financial, economic, political and military. Thus, we can not accept Mridula's contention that the Punjab experience was not somewhat exceptional or that in saying we are underplaying the importance of the colonial reality.

In this connection a brief reference may be made to the idea of "Punjab political tradition" popularized many years ago by Dungen. According to him at the heart of this tradition "lay the desire to create and preserve a stable rural base".[18] Dungen's approach is very narrow inasmch as he finds the expression of this desire only in the attempt by the British government to prevent land transfer from agriculturists to the professional moneylenders, thus ignoring the considerable benefits that accrued to the dominant sections of the Punjab rural society from canal construction and colonization programme. We would therefore suggest that if one has at all to speak of a "Punjab tradition", it should be understood in a wider sense to emphasize the exceptional nature of the Punjab situation as outlined above—a more interventionist role played by the foreign government in promoting agricultural development and the interest of the landowning classes, as well as the rich dividend that this involvement paid.

Finally, a brief reference may be made to the question why despite the fact that canal colonization provided vast resources Punjab agriculture failed to fare better than it did. Imran Ali has taken the view that "Important reasons for this lay in the political and social context in which growth in the canal colonies occurred." In particular he has emphasized the point that "the distribution of the colony lands was governed by the need of the British to consolidate their political position, to fulfil military requirements and to maintain an extractive system in order to finance their administration".[19] One could also refer to such factors as the government failure to evolve high yielding seeds for crops other than cotton, sugarcane and wheat, lack of adequate irrigation facilities in other parts of the province.[20] Again such other adverse factors—exploitation of the marginal farmers and tenants by the landlords, merchant-moneylenders, use of traditional tools and implements may also be mentioned in this connection.

There is no denying the fact that, in varying degrees, all these constituted real obstacles to a better performance by the Punjab agriculture. However, we would argue that the "external" factor mentioned earlier should be given more importance than it has received so far and this is true not only of Punjab but the whole subcontinent. Theoretical considerations as well as the historical experience of countries like Japan suggest that agricultural transformation and industrial development are interdependent. For while agriculture can make a decisive contribution to industrial-urban development, the latter creates a favourable environment for agricultural transformation through its impact on the product as well as labour, land, capital and credit market. As aptly pointed out by one commentator, "In general the process of economic growth contains as one of its major characteristics interdependence between the various sectors of the economy: the growth (or retardation) of a particular sector influences, and in turn influenced by, the growth (or retardation) of the other sectors. . . . Agriculture and industry would follow this general trend, but with a variety of patterns from one country to another and from one phase to another within a single country."[21] The Indian experience was different in that this sectoral interdependence did not develop within the country. For though increased external demand for primary produce, rise in prices and the development of a railway network led to an expansion of India's agricultural sector in the late nineteenth century, India did not herself industrialize on a significant scale. Consequently the critical pull factor was very weak or virtually absent.

We are not directly concerned with the question why industrialization efforts were so weak, but certainly one important factor was that state support which played critical role in the industrialization efforts in Japan, Germany, Canada and Australia was not forthcoming in the India situation. In other words, India's backwardness in the field of modern industrial developemnt was to a large extent due to the fact that the government followed a policy which was *laissez faire* in appearance but actually one of direct intervention in favour of British industries.[22] Anyway, as India failed to industrialize and direct government efforts for agricultural development remained extremely limited, after an initial spurt the agricultural sector began to stagnate with the onset of the twentieth century. This stagnation, in turn, became an obstacle to rapid industrialization. Such a state of affairs was inevitable because, the two sectors being interrelated, retardation in one was bound to affect the other. Thus, while looking for the "agrarian origin"[23] of industrial backwardness in India we should also take into account the experience in the late nineteenth century when agriculture expanded but there was little "support" from an emerging industrial sector. Incidentally, weak industrialization efforts during these earlier years in itself became an obstacle in the way of rapid development in this sector in the later period. For when industrialization efforts accelerated after the World War I the base was too narrow. As aptly pointed out by K.N. Chaudhuri, the highly favourable conditions which characterized Indian exports in the second half of the nineteenth century could have been turned to India's long-term advantage had there been systematic plans for economic modernization. But the opportunity was lost because Indian foreign trade was a dependent factor in an imperial system, which had different objectives than the transformation in the domestic economy.[24]

The fact that the canal colonies witnessed some growth does not negate the validity of our contention, in a way it underlines even more clearly the point that though significant growth in crop output could be possible in region with a special advantage in a certain area (greater scope for the extension of cultivation through government financed canal construction), overall transformation of the agrarian economy would remain a distant possibility in the absence of the development of the secondary sector.

NOTES

1. *Census of India,* 1921, *Punjab, Report,* Part 1, p. 2.
2. R.B. Buckley, *Irrigation Works in India and Egypt* (London, 1983), p. 110.
3. Thus, Kusum Nair's contention that the "Punjab cultivator appears to have been born to the ethic of utilizing irrigation for production, protection and profit" can not be taken without some reservation. For details see her *In Defense of the Irrational Peasant* (University of Chicago Press, 1979), p. 77.
4. R.C. Kennedy, Superintending Engineer, Irrigation Dept. of the Punjab, was of the opinion that "The present methods of distribution are most crude and out of date-unworthy of a department which considers itself scientific." For details see *Report of the Indian Irrigation Commission,* 1901-3, Vol. 4.
5. Krishna Bharadwaj, "A Note on Commercialization in Agriculture" in K.N. Raj *et al.* (eds.), *Essays on the Commercialization of Indian Agriculture* (Delhi, 1985), p. 338.
6. For details see Imran Ali, *The Punjab Under Imperialism* 1885-1947 (New Delhi, 1988), Ch. 6.
7. *Census of India,* 1931, *Punjab, Report,* part 2, Table XI.
8. According to one view members of the menial classes constituted 90 per cent of the subtenants. For details see Mark. K. Jurgensmeyer, "Political Hope: The Quest for Political Identity and Strategy", Part 1 (Unpublished Ph.D. Thesis, University of California, Berkely, 1974), p. 87.
9. Shyamala Bhatia, *Social Change and Politics in Punjab* 1898-1910 (New Delhi, 1987).
10. Prem Chowdhry, *Punjab Politics: The Role of Sir Chhotu Ram* (New Delhi, 1984), pp. 205-6.
11. Cited by *ibid.*, p. 289.
12. Richard G. Fox, *Lions of the Punjab* (University of California Press, 1985), p. 59.

13. *Ibid.* p. 59.
14. Mridula Mukherjee, "Commercialization and Agrarian Change in Pre-independence Punjab" in K.N. Raj *et al.* (eds.), *Essays on the Commercialization of Indian Agriculture* (Delhi, 1985), p. 53.
15. *Ibid.*, p. 73.
16. *Ibid.*, p. 54.
17. At the all-India level public canals irrigated only 15 per cent of the total cultivated area in 1938-9. In the same year only 4.4 million acres under paddy, 7.8 million acres under wheat, 6.2 million acres under cotton and 2.0 million acres under sugarcane (or about 10 per cent of the total cropped area) were sown with high-yielding varieties of seeds. For details see Vera Anstey, *Economic Development of India, New Impression* (London, 1957). Incidentally, the use of commercial fertiliser remained virtually unknown.
18. P.H.M. van den Dungen, *The Punjab Tradition* (London, 1972), p. 31.
19. Imran Ali, pp. 237-8.
20. The government paid little attention to the needs of the non-colony districts. The Deputy Commissioner of Gurgaon complained that the Punjab Agricultural Department was hardly a provincial agricultural department. It might best be described as "Canal colonies wheat and cotton improvement Department". For details see *Report of the Royal Commission on Agriculture in India,* Vol. VIII, *Evidence Taken in Punjab*, p. 64.
21. Kazushi Ohkawa, "Concurrent Growth of Agriculture and Industry: A Study of the Japanese Case" in R.N. Dixey (ed.), *International Explorations of Agricultural Economics* (Ames, Iowa, 1964), p. 205.
22. H.G.J. Aikten (ed.), *The State and Economic Growth* (New York, 1959).
23. Vasant Kaiwar, "The Agrarian Origin of India's Under Development", *South Asia Bulletin* 1, 2.
24. K.N. Chaudhuri, "Foreign Trade and Balance of Payments (1757-1947)" in Dharma Kumar (ed.), *The Cambridge Economic History of India,* Vol. 2 (Cambridge, 1984), p. 877. The same view had been taken by a large number of other scholars. See, for example, Arthur Lewis (ed.), *Tropical Development* 1880-1913: *Studies in Economic Progress* (London, 1970); R.S. Rungta, *Rise of Business Corporations in India,* 1851-1900 (Cambridge, 1970), p. 135; A.K. Bagchi, *The Presidency Banks and the Indian Economy,* 1876-1914 (Calcutta, 1989), p. 108.

APPENDICES

STATISTICAL TABLES

A. FINANCIAL STATEMENT ON THE MAJOR OR PRODUCTIVE WORKS (MILLION RS)

Years	*Accumulated Amount to the End of Year*				*Amount for the Year*			
	Capital	*Interest Charges*	*Net Revenue*	*Net Profit*	*Gross Revenue*	*Working Expenses*	*Net Revenue*	*% on Capital*
1887	57.6	3.2	50.9	19.1	4.0	1.7	2.3	4.0
1888	58.9	3.4	53.3	19.3	4.4	1.9	2.4	4.1
1889	60.6	3.6	56.2	19.8	4.9	2.1	2.8	4.7
1890	62.5	3.9	59.8	21.1	5.8	2.2	3.6	5.9
1891	67.0	4.1	63.6	22.5	6.0	2.2	3.8	6.7
1892	69.8	4.4	68.0	24.3	6.9	2.5	4.4	6.2
1893	73.5	4.7	70.9	44.3	5.4	2.4	2.9	5.3
1894	76.5	4.9	74.2	24.8	6.0	2.7	3.3	6.6
1895	80.2	5.2	78.8	26.3	7.1	2.8	4.2	5.3
1896	84.1	5.6	86.6	31.0	10.1	3.1	7.9	9.3
1897	87.0	5.9	95.6	36.7	12.1	3.2	9.0	10.4
1898	89.2	6.2	104.5	42.2	12.4	3.6	8.9	9.9
1899	90.4	6.6	113.8	48.0	13.4	4.1	9.3	10.2
1900	92.0	6.9	124.0	54.8	14.6	4.3	10.3	10.1
1901	80.4	6.9	131.5	61.6	14.6	4.3	10.3	11.5
1902	103.5	7.5	144.5	69.2	16.6	4.8	11.8	11.4
1903	106.9	8.0	158.3	77.9	18.6	5.7	12.9	12.0
1904	108.8	8.4	171.7	81.8	19.6	6.2	13.4	12.3
1905	110.2	8.9	183.7	95.1	13.8	6.6	11.7	10.7
1906	111.1	92.3	200.0	105.6	7.0	22.2	15.2	13.7
1907	112.4	96.0	211.0	115.0	7.2	20.3	13.1	11.6
1908	112.8	99.7	225.0	125.2	7.3	21.1	13.8	12.3
1909	113.3	103.3	239.4	136.0	7.9	22.4	14.5	12.8
1910	114.3	107.0	255.0	147.9	7.8	23.4	15.6	13.7
1911	116.2	110.8	273.7	162.9	8.4	27.1	18.8	16.2
1912	146.3	118.7	297.8	179.2	8.7	32.8	24.1	16.5
1913	167.8	125.4	323.8	198.4	8.8	34.3	25.9	15.5
1914	171.7	131.1	349.0	217.9	9.3	34.6	25.2	14.7
1915	217.0	144.1	373.7	228.6	10.9	34.6	23.7	10.9
1916	220.6	151.1	401.3	250.3	11.0	39.6	28.7	13.0
1917	223.3	158.5	428.0	269.5	11.1	37.8	26.7	12.0
1918	224.5	166.1	458.3	292.2	11.6	41.3	30.3	13.5

1919	224.5	173.6	490.7	317.0	12.4	44.8	32.4	14.4
1920	225.2	190.0	523.4	342.4	14.2	46.8	32.7	13.0
1921	227.0	188.4	556.9	368.5	16.9	50.5	33.5	14.8
1922	229.0	195.9	593.3	397.4	16.8	53.1	36.4	15.9
1923	232.1	203.6	633.5	429.9	55.6	15.4	40.2	15.9
1924	232.6	211.4	677.8	466.4	16.4	60.6	44.3	16.9
1925	231.5	218.2	724.1	505.9	15.7	63.2	47.6	20.5
1926	232.6	226.0	765.5	539.5	17.0	58.3	41.4	17.8
1927	305.7	327.7	781.6	573.9	18.5	64.7	46.1	15.1
1928	317.5	249.7	851.4	601.7	20.7	60.5	39.8	12.5
1929	323.4	261.7	885.7	624.0	24.5	68.4	44.0	13.6
1930	327.8	274.6	927.2	652.6	23.6	65.0	31.4	12.6
1931	331.8	287.7	970.1	782.4	15.4	58.4	43.0	13.0
1932	332.9	301.0	1012.0	710.6	16.9	58.4	41.4	12.5
1933	337.1	314.4	1017.2	745.7	17.0	65.6	48.6	14.4
1934	331.4	327.0	1107.9	780.9	16.9	64.6	47.8	14.4
1935	333.1	338.8	1155.4	816.6	17.0	64.6	47.5	14.3
1936	335.9	351.0	1204.3	853.2	16.8	65.6	48.9	14.6
1937	345.4	363.8	1256.0	892.3	15.4	67.1	52.8	15.0
1938	366.7	377.1	1306.0	928.9	14.6	64.6	50.0	13.6
1939	373.9	391.3	1361.2	969.8	14.2	69.4	55.2	14.8
1940	378.9	405.3	1419.3	1013.9	14.8	72.9	58.1	15.3
1941	384.8	419.3	1478.3	1059.0	15.6	74.6	59.1	15.4
1942	389.9	433.7	1544.3	1110.6	15.8	81.8	66.0	16.9
1943	393.5	448.3	1617.1	1168.9	18.5	91.4	72.8	18.5
1944	398.8	463.0	1690.8	1227.9	22.1	95.9	73.7	18.5
1945	409.9	478.0	1766.0	1288.1	17.3	102.6	75.2	18.3

Note: 1887 stands for 1887-8 and so on.

B. FINANCIAL STATEMENT OF MINOR IRRIGATION WORKS (MILLION RS)

Years	*Capital Outlay*	*Revenue*	*Working Expenses*	*Net Income*
1887	1437	1472	1101	370
1888	1437	1449	998	451
1889	1437	1379	881	498
1890	1453	1419	1063	356
1891	1521	1440	881	559
1892	1596	1447	901	546
1893	1635	1655	965	690
1894	1656	1562	1158	404
1895	1767	1667	1141	526
1896	1970	1568	1121	447
1897	2129	1858	1170	688
1898	2169	2018	1308	710
1899	2226	1883	1413	470
1900	2062	1863	1377	486
1901	2268	2227	1379	848
1902	1302	1739	1347	392
1903	1302	1470	869	600
1904	559	1674	1138	536
1905	559	1446	1012	435
1906	559	1587	936	652
1907	559	1584	973	611
1908	559	1379	1034	345
1909	559	1674	976	698
1910	559	1762	1016	746
1911	576	1600	941	658
1912	592	1666	878	788
1913	592	1818	910	907
1914	592	1881	933	947
1915	592	1849	905	944
1916	592	1961	825	1157
1917	592	2068	832	1236
1918	604	1748	946	802
1919	604	1610	1068	542
1920	607	1542	1218	323
1921	613	1413	1178	236
1922	613	1991	1239	752
1923	614	1077	494	582
1924	614	965	344	494
1925	3914	1864	1179	684
1926	3933	1639	1032	501
1927	3971	830	835	5
1928	3982	876	883	7
1929	5982	1151	1947	796
1930	5967	1036	1930	894
1931	5961	930	1309	379
1932	5961	1314	1247	67
1933	5959	1549	1373	176
1934	13760	1413	1477	64

1935	13739	1481	1342	139
1936	13349	1536	1359	177
1937	13227	1517	1342	176
1938	13194	1290	1351	61
1939	13182	1679	1153	526
1940	13232	1584	1202	382
1941	12980	1642	1157	486
1942	12951	1529	1290	239
1943	12985	1839	1576	263
1944	12997	1864	1811	53
1945	13054	1851	2100	250

Note: 1887 stands for 1887-8 and so on.

C. AREA IRRIGATED FROM PUBLIC CANALS ('000 ACRES)
(As per Administration Report of the Public Works Department, Irrigation Branch)

Years	*Major or Productive Works*			*Minor Works*			*Grand Total*
	Kharif	*Rabi*	*Total*	*Kharif*	*Rabi*	*Total*	
1887	646	777	1423	505	413	918	2341
1888	716	1034	1750	572	426	997	2747
1889	662	1202	1864	542	453	995	2858
1890	758	1276	2034	520	463	983	3016
1891	697	1447	2144	454	507	1049	3194
1892	832	1025	1848	491	589	1080	2928
1893	711	1193	1904	410	549	959	2862
1894	704	1152	1856	500	563	1063	2919
1895	986	1867	2825	478	461	939	3792
1896	1535	2114	3649	534	426	960	4609
1897	1778	2276	4053	566	593	1158	5212
1898	1414	2459	3873	453	531	984	4857
1899	1888	2497	4385	521	339	860	5244
1900	2354	2289	4643	670	688	1358	6001
1901	1686	2806	4492	508	573	982	5473
1902	2071	2793	4864	425	391	816	5680
1903	2451	3160	5611	316	463	780	6390
1904	2382	3307	5689	297	321	617	6306
1905	2294	3869	6163	298	453	751	6914
1906	2435	3468	5903	296	507	803	6706
1907	2226	3289	5515	258	266	525	6040
1908	2696	3690	6385	442	526	668	7353
1909	2286	3831	6117	426	544	970	7087
1910	2254	4016	6271	459	497	956	7227
1911	2755	4714	7470	403	460	862	8332
1912	2929	4532	7461	468	500	968	8429
1913	3070	4282	7352	465	507	972	8323
1914	3185	699	7884	466	613	1079	8963
1915	2887	5096	7983	363	540	902	8885
1916	3442	5052	8494	452	634	1086	9580
1917	3334	4653	7986	464	614	1078	9064
1918	3465	4883	8348	314	355	778	9016
1919	4216	5292	9508	455	493	948	10457
1920	4130	5376	9506	408	360	768	10274
1921	4278	6124	10402	368	476	843	11246
1922	4238	6202	10440	430	586	1016	11455
1923	4138	6226	10364	208	208	415	10779
1924	4457	5830	10287	209	293	503	10789
1925	4718	5758	10476	379	255	633	11109
1926	4716	6565	10481	357	319	676	11158
1927	4641	6057	10697	203	141	344	11042
1928	5445	6665	12110	221	137	358	12469
1929	5304	6584	11888	352	234	587	12575
1930	5282	6356	11638	328	303	631	12269
1931	4835	6076	10911	350	370	721	11632
1932	4720	5805	10524	364	351	715	11240

1933	5306	5962	11268	366	383	749	12017
1934	4933	5540	10473	324	352	676	11149
1935	5533	5572	11105	362	414	776	11881
1936	5745	6088	11833	410	404	815	12648
1937	6023	6224	12247	396	418	814	13061
1938	6059	6229	12289	413	473	885	13174
1939	6163	6518	12680	394	456	850	13531
1940	6206	6531	12737	404	446	451	13588
1941	6236	6514	12750	366	460	826	13576
1942	6311	7019	13331	367	459	826	14157
1943	6722	7132	13854	394	447	841	14695
1944	6833	7275	14108	343	419	762	14870
1945	6998	7927	14926	366	458	825	15750

Note: 1887 stands for 1887-8 and so forth.
No adjustment made for boundary changes in 1901.

D. AMOUNT COLLECTED AS WATER RATES AND WATER ADVANTAGE RATE AND ESTIMATED GROSS VALUE OF PRODUCE IN AREA IRRIGATED BY PUBLIC CANALS (MILLION RS AND THOUSAND ACRES)

Years	*Water Rate*	*Water Advantage Rate*	*Value of Crops*	*Area Assessed to Water Rates*
1887	3.0	0.46	68.5	2250
1888	3.4	0.46	75.2	2564
1889	4.0	0.45	81.4	2797
1890	4.6	0.46	88.2	2843
1891	5.0	0.25	89.5	3067
1892	6.0	0.14	71.5	2806
1893	4.5	0.22	74.8	2757
1894	4.9	0.31	74.3	2762
1895	5.9	0.43	93.2	3386
1896	9.3	0.53	150.8	4566
1897	10.3	0.58	149.1	5155
1898	10.3	0.68	132.9	4762
1899	11.3	0.78	185.2	5192
1900	12.4	0.98	156.1	5923
1901	12.0	1.47	147.2	5413
1902	13.9	1.50	157.5	5625
1903	15.2	1.82	163.6	6331
1904	15.6	2.21	166.4	6246
1905	14.9	1.97	167.5	6837
1906	17.9	2.31	162.0	6629
1907	15.9	2.37	157.8	5961
1908	16.7	2.49	180.2	6669
1909	17.0	3.05	206.8	6388
1910	17.7	3.46	207.0	6625
1911	19.8	5.83	258.3	7649
1912	23.3	8.07	307.0	7946
1913	23.7	10.10	264.5	7962
1914	23.3	10.18	277.6	8522
1915	22.8	10.79	275.0	8412
1916	26.6	11.93	320.9	8890
1917	26.0	10.68	333.7	8230
1918	26.1	13.60	550.9	8430
1919	30.9	12.51	624.1	9948
1920	31.6	13.49	516.8	9418
1921	30.9	17.89	672.1	10562
1922	34.1	17.12	494.2	10696
1923	34.1	20.04	550.5	10099
1924	38.0	20.91	577.1	10143
1925	41.3	20.23	545.3	10274
1926	38.5	18.12	457.8	9972
1927	40.1	23.02	447.3	9853
1928	39.1	19.65	550.6	11869
1929	43.7	27.43	492.7	12006
1930	40.4	22.95	294.5	11584
1931	38.1	18.30	327.4	11245
1932	37.1	20.10	360.5	11181

1933	41.9	21.39	322.7	10983
1934	38.9	24.49	341.6	10235
1935	39.6	23.77	389.9	10981
1936	41.0	23.22	474.1	11629
1937	41.5	24.08	403.2	12221
1938	39.9	22.72	389.0	11890
1939	41.6	25.20	507.5	12660
1940	43.9	26.79	441.2	12633
1941	44.4	28.00	648.0	12757
1942	45.9	32.98	1249.1	13160
1943	49.5	38.78	1166.4	13700
1944	49.6	41.38	1258.1	13820
1945	51.2	48.26	1442.3	14579

Note: 1887 stand for 1887-8 and so on.

E. AREA IRRIGATED FROM DIFFERENT SOURCES ('000 ACRES)
(As per Season and Crop Reports, Punjab)

Years	*Public Canals*	*Wells*	*Misc. Sources*	*Total Irrigated Area*	*Total Sown Area*
1887	2006	3889	1145	6991	22790
1888	2333	3883	1163	7379	23470
1889	2476	3959	1052	7487	21231
1890	2515	3822	1124	7461	23536
1891	2606	3696	1047	7348	21409
1892	2550	3429	1087	7066	26732
1893	2523	3450	1094	7067	25646
1894	2564	3072	1074	6714	24480
1895	3161	3731	953	7827	19054
1896	3733	4007	980	8720	18516
1897	4395	3971	1106	9472	25810
1898	3982	3957	1029	8967	20739
1899	4244	4155	978	9376	14994
1900	5192	3018	1236	9446	28570
1901	4531	3747	574	8851	23212
1902	4888	3827	639	9353	25527
1903	5509	3452	718	9679	27792
1904	5375	3522	623	9520	26269
1905	5606	3696	675	9977	26651
1906	6801	3132	825	10758	29875
1907	5941	3989	597	10527	24529
1908	6406	3030	738	10174	29298
1909	6242	2986	716	9943	29648
1910	6393	3071	630	10094	28597
1911	6964	3420	570	10955	26309
1912	7026	3602	674	11302	27510
1913	7059	3761	650	11470	27329
1914	7594	3213	689	11495	31716
1915	7444	3605	627	11677	25904
1916	8072	3433	691	12196	31704
1917	6782	2951	681	11314	32858
1918	7531	3829	514	11874	21854
1919	8704	3536	688	12929	29111
1920	8495	3876	583	12954	24561
1921	4997	3587	709	13293	31026
1922	9629	3209	673	13511	31789
1923	9294	3214	663	13170	30606
1924	9272	3192	619	13083	31721
1925	9499	3715	605	13819	29710
1926	9586	3484	609	13679	30407
1927	9352	3916	606	13873	29473
1928	10382	4122	561	15065	31926
1929	10649	4819	576	16043	30954
1930	10239	4014	561	14814	30265
1931	9929	3767	571	14267	32007
1932	9479	4174	595	14247	30058
1933	10216	3549	638	14403	34494

1934	9529	4352	545	14425	29834
1935	10143	4288	587	15018	31851
1936	10886	4087	632	15605	32643
1937	11203	4346	591	16140	31572
1938	11166	4749	629	16544	28845
1939	11406	4722	640	16768	29947
1940	11565	4681	652	16898	32803
1941	11639	4618	658	16915	32299
1942	12110	3862	645	16617	34989
1943	12483	4216	683	17382	33003
1944	12647	4093	595	17335	34667
1945	13388	3996	671	18055	35956
1946	12780	4190	601	17571	32699

Note: 1887 stand for 1887-8 and so on.

F. CROPS IRRIGATED AND UNIRRIGATED ('000 ACRES)

Years	*Wheat*		*Gram*		*Maize*		*Barley*		*Jowar*	
	a	*b*	*a*	*b*	*a*	*b*	*a*	*b*	*a*	*b*
1887	2773	4904	210	3291	541	907	532	1116	472	2022
1888	3049	5489	247	3021	553	926	561	1419	447	2167
1889	3175	4140	297	1810	564	945	480	710	425	2469
1890	3081	5599	317	2546	515	969	504	1554	393	2409
1891	3059	4709	418	2905	572	885	503	880	378	2087
1892	2781	5864	354	3721	450	1119	484	1451	476	3654
1893	3048	6670	273	3635	581	928	429	1463	267	2348
1894	2817	6477	237	3433	482	857	325	1166	307	2380
1895	3204	3891	328	1597	566	987	408	506	425	1696
1896	3269	3523	370	1064	601	987	626	656	559	1530
1897	3675	5197	462	2584	736	1129	633	1315	429	2399
1898	3767	4175	355	1428	706	1111	537	625	308	1080
1899	3768	2185	281	480	688	701	652	313	334	321
1900	3575	6235	408	3806	600	1125	460	337	432	1856
1901	4066	2605	420	1102	644	846	445	230	351	727
1902	4036	3694	529	2049	680	964	413	484	389	1168
1903	4399	5168	634	3046	597	867	405	780	356	1788
1904	4273	5141	448	2776	575	760	343	1199	385	1095
1905	4856	5532	686	3194	511	584	456	1081	370	485
1906	4441	5206	752	3748	499	697	309	1026	386	1172
1907	4374	3019	507	1262	571	597	516	844	336	1033
1908	4054	4343	659	3569	449	780	352	829	275	1352
1909	4128	4548	706	4039	465	784	297	797	248	1067
1910	4359	4526	646	3848	448	759	254	750	266	1077
1911	4551	5174	683	3417	451	504	324	1015	293	261
1912	4695	4072	699	2702	490	608	342	665	271	942
1913	4506	3966	621	2229	471	612	428	541	271	977
1914	4567	5339	895	4299	455	592	310	998	228	1048
1115	4774	4218	862	2804	579	605	474	566	566	692
1916	4935	4533	900	4216	534	737	330	823	290	1204
1917	4658	5267	776	5229	295	924	269	1206	143	683
1918	4740	2944	646	1403	580	571	493	372	182	471
1919	4877	3936	819	3306	525	630	369	835	202	818
1920	4993	2759	815	1413	479	585	391	240	230	692
1921	4905	3884	1181	3967	518	594	395	717	266	947
1922	5049	4571	1174	4254	513	610	328	845	207	744
1923	5005	4666	956	3245	452	599	306	940	191	794
1924	4728	4929	944	4753	379	543	246	690	219	836
1925	5100	4382	909	2801	379	552	301	503	167	756
1926	5046	4334	927	3758	378	598	262	506	181	802
1927	5129	3895	970	3119	495	591	312	523	185	840
1928	5333	4636	1048	3111	482	566	391	949	239	710
1929	5591	4301	1104	2047	539	603	432	489	238	870
1930	5356	3931	1243	2878	519	576	256	400	192	698
1931	4917	4163	1383	4164	427	577	197	433	211	803
1932	4972	3619	1175	2719	474	560	250	368	205	908
1933	4973	4800	1244	5342	341	715	192	517	176	700
1934	5220	3818	1001	2620	540	595	253	359	163	680

1935	5190	4111	1007	3701	533	558	226	440	173	649
1936	5151	4234	1205	3703	487	591	239	500	185	743
1937	5568	4377	1038	2716	535	568	297	498	180	658
1938	5820	3706	936	1366	542	568	335	240	202	676
1939	5797	3769	990	1423	554	589	374	356	203	575
1940	5811	4074	1061	2390	545	600	369	430	208	669
1941	5813	4195	1033	2428	599	589	345	459	211	562
1942	5739	4723	1216	3618	577	725	317	626	173	715
1943	5976	4013	1260	2842	678	649	332	428	174	660
1944	5917	4507	1251	3694	712	637	281	571	199	681
1945	6066	4461	1696	4454	711	615	248	492	192	626

Note: 1887 stands for 1887-8 and so on, (a) Irrigated, (b) Unirrigated.

G. CROPS IRRIGATED AND UNIRRIGATED

Years	*Rice*		*Bajra*		*Cotton*		*Sugarcane*	
	a	*b*	*a*	*b*	*a*	*b*	*a*	*b*
1887	388	434	136	2601	423	280	276	115
1888	391	381	148	2274	426	428	265	161
1889	440	403	114	2741	500	608	221	132
1890	444	316	133	2503	527	406	246	99
1891	416	389	92	1800	329	219	259	133
1892	400	298	161	3133	425	156	237	98
1893	456	458	91	2121	465	540	227	128
1894	443	522	74	2115	435	629	239	129
1895	420	416	68	1490	624	546	266	152
1896	412	178	111	1367	736	329	314	107
1897	513	347	198	2973	709	361	280	86
1898	550	290	160	1774	485	275	313	104
1899	412	89	181	605	630	133	289	92
1900	505	361	334	3423	806	328	281	89
1901	497	301	225	1571	692	296	312	94
1902	451	260	272	2378	846	357	290	72
1903	494	318	271	2870	939	246	289	88
1904	424	250	195	2197	1060	442	281	100
1905	434	125	173	923	572	216	120	62
1906	494	233	221	2280	894	366	214	65
1907	441	255	197	2432	973	345	316	78
1908	483	238	257	2968	1087	361	292	77
1909	543	256	210	2584	806	471	328	89
1910	498	215	156	2256	916	333	311	89
1911	424	90	126	1029	1188	274	216	82
1912	514	227	219	2458	1137	306	281	86
1913	582	219	204	2625	1379	448	321	90
1914	554	240	215	2523	1302	386	290	76
1915	586	160	235	1673	7664	163	282	66
1916	787	277	283	2750	872	192	350	64
1917	726	278	199	2349	2667	376	413	90
1918	596	1412	224	1337	1272	146	355	119
1919	745	218	260	2416	1836	235	388	94
1920	711	210	228	2194	1717	240	279	78
1921	626	196	417	2906	1054	95	326	18
1922	736	193	387	2732	1153	120	436	60
1923	691	194	306	2545	1543	207	404	79
1924	592	188	289	2296	2059	267	310	85
1925	743	226	258	2305	2421	281	302	88
1926	645	239	266	2426	2271	253	362	86
1927	722	219	360	2357	1711	130	415	83
1928	708	176	352	2127	2378	131	326	75
1929	811	164	501	2863	2089	120	256	50
1930	804	172	443	2793	2023	141	362	63
1931	652	148	336	2897	2015	145	404	71
1932	692	187	389	3014	1759	131	467	91
1933	778	257	344	3012	2228	221	323	83
1934	708	248	337	2706	2160	187	380	82

1935	720	252	322	2697	2586	217	390	85
1936	799	241	336	2515	2678	231	464	91
1937	869	217	327	2289	2885	250	419	91
1938	837	193	402	2239	2760	143	288	66
1939	782	195	554	2507	2466	175	352	65
1940	769	182	655	3208	2504	165	467	82
1941	751	141	655	3050	2634	167	382	76
1942	890	210	835	3574	2171	148	367	81
1943	1047	209	688	2995	2467	135	464	88
1944	1063	202	708	2882	2273	149	496	106
1945	1118	203	708	2770	2234	108	484	86

Note: 1887 stands for 1887-8 and so on, (a) Irrigated, (b) Unirrigated.

H. AREA UNDER CEREALS + PULSES, FODDER CROPS, AREA FAILED AND DOUBLE-CROPPED ('000 ACRES)

Years	Cereals and Pulses		Fodder Crops		Failed Area		Double-cropped Area	
	a	b	a	b	a	b	a	b
1887	5319	18033	N.A.	N.A	N.A.	N.A.	N.A.	N.A.
1888	5628	18056	N.A.	N.A.	N.A.	N.A.	N.A.	N.A.
1889	5692	15605	N.A.	N.A.	N.A.	N.A.	N.A.	N.A.
1890	5591	18331	N.A.	N.A.	N.A.	N.A.	N.A.	N.A.
1891	5668	16043	N.A.	N.A.	N.A.	N.A.	N.A.	N.A.
1892	5371	22412	350	425	N.A.	N.A.	N.A.	N.A.
1893	5354	20939	335	321	N.A.	N.A.	N.A.	N.A.
1894	4869	19946	358	280	N.A.	N.A.	N.A.	N.A.
1895	5638	12621	467	147	N.A.	N.A.	N.A.	N.A.
1896	6278	10940	536	176	733	5670	N.A.	N.A.
1897	7042	18344	660	406	503	3246	N.A.	N.A.
1898	6625	12567	739	915	740	5103	N.A.	N.A.
1899	6626	5649	980	553	1187	7095	N.A.	N.A.
1900	6618	20585	867	779	509	2671	N.A.	N.A.
1901	6856	8897	780	773	1143	6424	N.A.	N.A.
1902	7027	12798	937	885	712	5074	N.A.	N.A.
1903	7368	17044	867	1218	584	3337	N.A.	N.A.
1904	6791	15059	1199	1055	687	3626	N.A.	N.A.
1905	7754	13369	1273	835	493	4891	N.A.	N.A.
1906	7332	16106	1273	1229	804	3126	234	49
1907	7169	11017	1321	1569	1462	6687	266	27
1908	6761	15867	1232	1467	1141	3920	258	43
1909	6808	16052	1211	1597	857	3066	272	28
1910	6810	15220	1290	1634	743	2750	263	39
1911	7044	12334	1878	1018	1012	3812	362	53
1912	7430	12975	1878	1772	765	4061	288	27
1913	7275	12406	1932	1782	665	3791	308	37
1914	7450	16514	1950	1949	680	2909	294	35
1915	8066	11754	2177	1001	884	5640	382	25
1916	8317	15942	2043	2013	750	3458	326	51
1917	7261	17483	1828	2184	994	3621	235	32
1918	7694	7940	2154	932	756	3993	318	22
1919	8004	13237	2178	1595	672	3076	286	21
1920	8056	9122	2323	1474	672	3076	286	21
1921	8597	14539	2574	1519	800	4579	314	17
1922	8660	15078	2548	1683	706	2570	289	28
1923	8143	14203	2259	1853	730	3145	282	25
1924	7577	15409	2370	1648	840	4114	292	22
1925	8045	12785	2452	1821	1005	4095	269	29
1926	7918	13765	2448	1706	1003	3224	262	17
1927	8441	12646	2601	1755	976	3668	308	18
1928	8814	13280	2795	1684	1271	6393	318	23
1929	9502	12506	2716	1784	1089	5314	264	29
1930	9071	1284	2592	1828	1097	4045	258	21
1931	8368	14448	2646	1826	1134	4134	279	19
1932	8421	12576	2770	1815	1120	5495	270	30

1933	8368	16594	2788	1970	1043	4881	254	21
1934	8454	12207	2865	1977	831	3667	260	24
1935	8442	13503	3031	2037	955	4229	257	25
1936	8674	13710	3134	1856	863	3777	257	25
1937	9061	12351	3272	1754	839	4614	267	21
1938	9348	9853	3509	1365	1286	4990	291	22
1939	9572	10327	3640	1406	839	4799	279	27
1940	9731	12530	3509	1707	856	4372	262	19
1941	9794	12344	3442	1541	599	4435	290	22
1942	10180	15311	3303	1685	686	2610	243	23
1943	10563	12836	3403	1720	645	3142	221	22
1944	10534	14191	3419	1856	748	3146	220	18
1945	10580	12470	3718	1936	—	—	—	—

Notes: (a) Irrigated, (b) Unirrigated, N.A. Not available. 1887 stands for 1887-8 and so no.

GLOSSARY

abadi	village site.
abadkar	peasant grantee.
abiana	water rate or occupier's rate levied for canal-irrigation.
abpashi	cultivation through irrigation.
arthi	commission agent.
arthri	agricultural labourer, generally of low caste origin, receiving payment in kind.
bania	Hindu shopkeeper-moneylender.
bar	tableland lying between two rivers.
barani	watered by rains only.
batai	division of produce system of taking land revenue/rent in kind.
begar	forced labour.
benami	transfer of land in the name of a fictitious person.
chahi	watered from well.
chak	blocks of land grouped together for administration.
chamar	worker in leather.
chari	fodder crop.
chuhra	low-caste sweeper.
deshi	indigenous
doab	tract of land between rivers.
fasal	crop.
girdawar	a revenue official.
haq abpasi	area proposed for irrigation (popular and erroneous designation).
hithar	riverine lowland.
ilaqad	estate; locality.
jajmani	relationship between service and landholding castes.
jamabandi	village record of rights.
Jangli	indigenous residents of the colonies.
Jhallar	lift for irrigation.
jinswar	crop statements.
kamin	village menial.
kanungo	revenue official who supervise the work of the *patwaris*.
khadir	riverain tract.
khal	water-course.
kharaba	failed area of crops.
kharif	summer or autumn harvest.
kiari	division of the field for irrigation.
killa	one-twenty-fifth of a colony square of little over an acre.
killabandi	demarcation of land in *killas*.

kishan	peasant, cultivator.
lambardar	village headman.
malik	owner.
malikana	fee paid in recognition of proprietory right.
mandi	market.
maund	a unit of weight, little over 80 lbs.
milan khasra	annual area statement.
nahri	watered from canals.
nai	barber.
nazrana	payment made to the government on acquisition of a colony grant.
patti	subdivision of an estate.
pattidar	village shareholder.
patwari	village accountant or record keeper.
rabi	winter or spring harvest.
rais	capitalist grantee.
rajbaha	major distributary of a canal.
sahukar	professional moneylender, trader.
sanji	an agricultural labourer.
sufedposh	a Yeoman grantee.
sailab	irrigation by river flood.
sepidar	one who rendered the service called *sep* to the landowners and received payment in kind at the end of the harvest.
sepidari	relationship between the services and landholding castes.
shamilat-deh	common land of the village.
tahsil	section or revenue and administrative subdivision, of a district.
tarkhan	carpenter.
warabandi	system by which the supply of water to canal branches, distributaries, water courses was regulated in rotation.
warashikni	infringement of *warabandi* schemes by irrigators; taking water out of turn.
zamindar	an owner or occupier of land.

SELECT BIBLIOGRAPHY

Government Publications

Annual Reports

Report on the Administration of the Canal Colonies (1892-1940).

———. *Administration of the Public Works Department, Irrigation Branch,* Punjab, 1887-8–1945-6.

———. *Indian Central Cotton Committee,* 1945.

———. *Internal Trade of the Punjab by Rail and River,* 1882-3–1921-2.

———. *Operations of the Department of Land Records and Agriculture, Punjab,* 1893-4–1904-5.

———. *Operations of the Department of Agriculture, Punjab,* 1906-7–1940-1.

———. *Revenue (*later *Land Revenue) Administration, Punjab,* 1887-8–1945-6.

———. *Season and Crops, Punjab,* 1906-7–1945-6.

Assessment Reports (AR)

AR of the Jhelum Canal (Shahpur), 1909.

———. *Gugera Branch II Lower Chenab Canal,* 1912.

———. *Gujranwalla Tahsil, Gujranwalla District,* 1912.

———. *Fazilka Tahsil, Ferozepore District,* 1913.

———. *Mamdot Jagir, Ferozepore District ,* 1912-13.

———. *Garhshankar Tahsil, Hoshiarpur District,* 1913.

———. *Rakh Branch Circle, Chenab Colony,* 1913.

———. *Kharian Tahail, Gujrat District,* 1913-14.

———. *Una Tahsil, Hoshiarpur District,* 1914.

———. *Gujrat Tahsil, Gujrat District,* 1914.

———. *Phalia Tahsil, Gujrat District,* 1915.

———. *Multan and Shujabad Tahsil, Multan District,* 1920.

———. *Jhang Branch Circles, Lyallpur District,* 1921.

———. *Jhang Tahsil, Jhang District,* 1925.

———. *Hafizabad Tahsil, Gujranwalla District,* 1925.

———. *Attock Tahsil, Attock District,* 1925.

———. *Montgomery Tahsil, Montgomery District,* 1934.

———. *Jaranwalla Tahsil, Lyallpur District,* 1936.

District Gazetteers (arranged chronologically)

Multan, 1901-2 (Lahore, 1902), 1923-4 (Lahore, 1925).

Hoshiarpur, 1904 (Lahore, 1906).

Jhelum, 1904 (Lahore, 1907).
Ludhiana, 1904 (Lahore, 1907).
Rawalpindi, 1906 (Lahore, 1909).
Gurgaon, 1910 (Lahore, 1911).
Rohtak, 1910 (Lahore, 1911).
Amritsar, 1914 (Lahore, 1914).
Gurdaspur, 1914 (Lahore, 1915).
Mianwali, 1915 (Lahore, 1915).
Ferozepore, 1915 (Lahore, 1916).
Hissar, 1915 (Lahore, 1916).
Lahore, 1916 (Lahore, 1916).
Shahpur, 1917 (Lahore, 1918).
Karnal, 1918 (Lahore, 1919).
Sialkot, 1920 (Lahore, 1921).
Gujrat, 1921 (Lahore, 1921).
Ambala, 1924 (Lahore, 1925).
Kangra, 1925 (Lahore, 1926).
Muzaffargarh, 1929 (Lahore, 1930).
Jhang, 1929 (Lahore, 1930).
Attock, 1930 (Lahore, 1932).
Montgomery, 1933 (Lahore, 1935).
Gujranwalla, 1935 (Lahore, 1936).

Settlement Reports (arranged chronologically)

Final Report on the Revision of Settlement of Attock District (Lahore, 1909).
———. *Revision of Settlement of Rawalpindi District* (Lahore, 1909).
———. *Third Regular Settlement of Rohtak District*, 1905-10 (Lahore, 1910).
———. *Second Revised Settlement of the Ludhiana District*, 1908-11 (Lahore, 1912).
———. *Third Regular Settlement of the Mianwali District* (Lahore, 1912).
———. *Settlement of Karnal District* (Lahore, 1912).
———. *Fourth Regular Settlement of Amritsar District*, 1910-14 (Lahore, 1914).
———. *Revised Settlement of the Frozepore District*, 1910-14 (Lahore, 1915)
———. *Second Revised Settlement of the Hoshiarpur District*, 1910-14 (Lahore, 1915).
———. *Chenal Colony Settlement* (Lahore, 1915)
———. *Fourth Regular Settlement of the Lahore District* (Lahore, 1916).
———. *Fourth Regular Settlement of the Gujrat District*, 1912-16 (Lahore, 1917).
———. *Fourth Regular Settlement of the Sialkot District* (Lahore, 1917).
———. *Settlement of the Multan District* (Lahore, 1921).
———. *Second Revised Settlement of the Ambala District*, 1915-20 (Lahore, 1921).
———. *Settlement of the Jhang and Gugera Branch Circles of the Layallpur District*, 1920-4 (Lahore, 1924).
———. *Second Revised Settlement of the Dera Gazi Khan District* (Lahore, 1925).
———. *Third Revised Settlement of the Pakpattan and Dipalpur Tahsils of the Montgomery District*, 1919-22 (Lahore, 1925).
———. *Fourth Revised Settlement of the Gujranwalla District*, 1923-7 (Lahore, 1927).
———. *Settlement of the Shaikhpura District* (Lahore, 1927).
——— *Fourth Regular Settlement of the Attock District* (Lahore, 1928).
———. *Settlement of the Muzaffargarh District*, 1920-25 (Lahore, 1929).

———. *Settlement of the Canal Irrigated Tracts of the Gujrat District* (Lahore, 1930).
———. *Settlement of the Lower Bari Doab Colony*, 1927-35 (Lahore, 1936).
———. *Settlement of the Lyallpur District* (Lahore, 1940).
———. *Settlement of the Gurgaon District*, 1938-43 (Lahore, 1944).

Other Reports

Census of India, 1901, *Punjab and Delhi, Report*, Vol. XII.
———. 1911, *Punjab and Delhi, Report*, Vol. XIV.
———. 1921, *Punjab and Delhi, Report*, Vol. XIV.
———. 1931, *Punjab, Report*, Vol. XVIIII.
———. 1941, *Punjab, Tables.*
Report of the Indian Irrigation Commission, 1901-3 (London, 1903).
———. *Punjab Colonies Committee*, 1907-8 (Lahore, 1908).
———. *Indian Cotton Committee*, Vol. 1 (Calcutta, 1919).
———. *Waterlogging in the Punjab* (Lahore, 1924).
———. *Royal Commission on Agriculture in India*, Vol. VIII: *Evidence Taken in the Punjab* (Calcutta, 1927).
———. *Punjab Provincial Banking Enquiry Committee*, 2 vols. (Lahore, 1930).
———. *Abiana Committee* (Lahore, 1934).
———. *Punjab Land Revenue Committee* (Lahore, 1937).
———. *Marketing of Wheat in the Punjab* (Lahore, 1939).
———. *Working of the Debt Conciliation Boards upto 31 December 1939* (Lahore, 1942).
———. *Punjab Live-Stock Census*, 1940 (Lahore, 1942).
———. *Famine Inquiry Commission* (Madras, 1945).
———. *Alienation of Land Inquiry*, West Pakistan (Lahore, 1949).

Miscellaneous Publications

Notes on Land Transfer and Agricultural Indebtedness in India (Calcutta, 1895).
The Punjab Land Administration Manual (Lahore, 1908).
The Punjab Settlement Manual (Lahore, 1915).
The Punjab Colony Manual (Lahore, 1922).
First-Third Regular Wages Survey of the Punjab, 1912, 1917, and 1922 (Lahore, 1913, 1918, 1923).
Punjab Irrigation Branch, Paper No. 2: "Waterlogging of the Soil in the Vicinity of Punjab Irrigation Canals and Measures for the Prevention of Water-logging".
Punjab Irrigation Branch, Paper No. 12: "On the Distribution of Water by Measurement".
Punjab Irrigation Branch, Paper No. 31: "An Estimate of Area Affected by, and Threatened with waterlogging with their past History and the Extent to which Remedial Measures will be Necessary".
Proceedings of Waterlogging Conferences, 1928-38.
Proceedings of Meeting of Waterlogging Board, Punjab, Vols. 1-5.

PUBLICATIONS OF THE PUNJAB BOARD OF ECONOMY INQUIRY

The Size and Distribution of Agricultural Holdings in the Punjab by H. Calvert (Lahore, 1920).
Sixty Years of Punjab Food Prices, 1861-1920 by W.H. Myles (Lahore, 1925).
Eighty Years of Punjab Food Prices (Lahore, 1926).

An Inquiry into Mortgages of Agricultural Land in Pothwar Assessment Circle of Rawalpindi District (Lahore, 1926).
Some Aspects of Batai Cultivation in the Layallpur District of the Punjab by H.R. Stewart (Lahore, 1926).
Accounts of Different System of Farming in the Canal Colonies of the Punjab by H.R. Stewart and Kartar Singh (Lahore, 1927).
The Size and Distribution of Cultivators Holdings in the Punjab by H. Calvert (Lahore, 1928).
A Note on the Sale of Land Between the Notified Agricultural Tribes in Punjab During the Quinquennium, 1922-3 to 1926-7 by C.P.K. Fazal (Lahore, 1931).
Finance and Marketing of Cultivators' Wheat in the Punjab by S.Kartar Singh (Lahore, 1934).
Studies in the Cost of Production of Crops in the Punjab (Lahore, 1934).
Market Practices in the Punjab by H.Calvert (Lahore, 1934).
A Summary of the More Important Results Arrived at or Indicated by the Agricultural Stations and Research Officers in the Punjab During 1930-1 to 1934-5 (Lahore, 1936).
Some Factors Affecting the Price of Wheat in the Punjab (Lahore, 1937).
Report on Factors Affecting Irrigation in the Punjab from Tubewells by A.M.R. Montague (Lahore, 1938).
Sales of Land in South-West Punjab, 1931-2 to 1933-4 (Lahore, 1938).
Guide to Punjab Government: Reports and Statistics (Lahore, 1939).
Cost of Production of Crops on a Canal Irrigated Estate in the Punjab, 1935-6 to 1939-40 by L. Singh and A. Singh (Lahore, 1944).
Agricultural Statistics in Punjab, 1901-2 to 1935-6, Part-1 (Lahore, 1937), Part 2, 1936-7 to 1943-4 (Lahore, 1945).
Rise in the Prices of Bhusa in the Punjab (Lahore, 1949).
Agricultural Statistics of the Punjab, Pakistan, 1901-2 to 1946-7 (Lahore, 1950).
Farm Accounts in the Punjab, 1926-7 to 1942-3.
An Economic Survey of Bhadas: Gaggar Bhana: A Village in Amritsar District (Lahore, 1928).
———. *Tehong: A Village in Jullunder District* (Lahore, 1931).
———. *Gijhi, A Village in Rohtak District* (Lahore, 1932).
———. *Kala Gaddi Thamman: A Village in Lyallpur District* (Lahore, 1934).
———. *Gajju Chak: A Village in Gujranwalla District* (Lahore, 1934).
———. *Bambu Sandila: A Village in Muzaffargarh District* (Lahore, 1935).
———. *Bhadas: A Village in Gurgaon District* (Lahore, 1936).
———. *Suner: A Village in Ferozepore District* (Lahore, 1936).
———. *Durrana Langana: A Village in Multan District* (Lahore, 1938).

Secondary Work: Books and Articles

Adams, John and R.C. West. "Money, Prices and Economic Development of India", *Journal of Economic History,* 34 (1979).
Agnihotri, H.L. and S.N. Malik. *A Profile in Courage: A Biography of Choudhry Chhotu Ram* (New Delhi, 1978).
Aguilar, F.V. "The Philippine Peasant as Capitalist: Beyond the Categories of Ideal Typical Capitalism", *Journal of Peasant Studies,* 17 (1989).
Aitken, H.G.J. (ed.), *The State and Economic Growth* (New York, 1959).
Alavi, Hamza. "India: Transition from Feudalism to Capitalism", *Journal of Contemporary Asia,* 10 (1980).
Ali, Imran. *Punjab under Imperialism,* 1885-1947 (Princeton, 1988).
———. "Malign Growth? Agricultural Colonization and the Roots of Backwardness in the Punjab", *Past and Present,* 114 (1987).

Ambirajan, S. *Classical Political Economy and British Policy in India* (Cambridge, 1978).

Amin, S. *Accumulation on a World Scale* (New York, 1978).

Ansari, Nasim. *Economics of Irrigation Rates: A Study in Punjab and Uttar Pradesh* (London, 1968).

Anstey, Vera. *The Economic Development of India*, New impression (London, 1957).

Bagchi, A.K. "Land Tax, Property Rights and Peasant Insecurity in Colonial India", *Journal of Peasant Studies*, 20 (1).

———. *The Political Economy of Underdevelopment* (Cambridge, 1982).

Banga, Indu. *Agrarian System of the Sikhs: Late Eighteenth and Early Nineteenth Century* (New Delhi, 1978).

Banerjee, Himadri. *Agrarian Society of the Punjab*, 1849-1901 (New Delhi, 1982).

Benerji, A.K. *Aspects of Indo-British Economic Relations*, 1858-1898 (New Delhi, 1982).

Baran, P.A. *The Political Economy of Growth* (New Delhi, 1962).

Barrier, N.G. "The Formulation and Enactment of the Punjab Alienation of Land Bill", *Indian Economic and Social History Review*, 2 (1965).

———. *The Punjab Alienation of Land Bill*, 1900 (Durham, N.C., 1966).

———. "The Punjab Disturbances of 1907: The Response of the British Government in India to Agrarian Unrest", *Modern Asian Studies*, 1940.

Bernstein, H. "Concepts for the Analysis of Contemporary Peasantries", *Journal of Peasant Studies*, VI (4).

Bhaduri, A. "A Study in Agricultural Backwardness Under Semi-Feudalism", *Economic Journal*, March 1973.

———. "Class Relations and Pattern of Accumulation in an Agrarian Economy", *Cambridge Journal of Economics*, March 1981.

———. *The Economic Structure of Backward Agriculture* (London, 1983).

Bharadwaj, Krishna. *Production Conditions in Indian Agriculture*: A Study Board on Farm Management Surveys (Cambridge, 1974). "A Note on Commercialization in Agriculture" in K.N. Raj *et al.* (eds.), *Essays on the Commercialization of Indian Agriculture* (Delhi, 1985).

Bhatia, B.M. *Famines in India: A Study of Some Aspects of the Economy of India* (New Delhi, 1964).

Bhatia, Shyamala. *Social Change and Politics in Punjab*, 1898-1910 (New Delhi, 1987).

Bhattacharya, Neeladri. "The Logic of Tenancy Cultivation: Central and South-East Punjab, 1870-1935", *Indian Economic and Social History Review*, 20 (1983).

———. "Agricultural Labour and Production: Central and South-East Punjab, 1870-1940" in K.N. Raj *et al.* (eds.), *Essays on the Commercialization of Indian Agriculture* (Delhi, 1985).

———. "Lenders and Debtors: Punjab Countryside: 1880-1940", *Studies in History*, 1, No. 2 (1985).

Bhattacharya, S. "Laissez Faire in India", *Indian Economic and Social History Review*, 1 (1965).

Black, R.D.C. *Economic Thought and the Irish Question* (Cambridge, 1959).

Blyn, George. *Agricultural Trends in India*, 1891-1947 (Philadelphia, 1966).

Buckley, R.B. *Irrigation Works in India and Egypt* (London, 1893).

———. *The Irrigation Works of India* (London, 1905).

Byres, T.J. "Land Reform, Industrialization and the Marketed Surplus in India: An Essay on the Power of Rural Bias", David Lehman (ed.), *Agrarian Reform and Agrarian Reformism* (London, 1974).

Calvert, H. *The Wealth and Welfare of the Punjab* (Lahore, 1922).

Chandra, B. *The Rise and Growth of Economic Nationalism in India* (New Delhi, 1966).

Charlesworth, Neil. *British Rule and the Indian Economy*, 1800-1914 (London, 1982).

———. "Rich Peasants and Poor Peasants in Late Nineteenth-Century Maharashtra", in Clive J. Dewey and N. Hopkins (eds.), *The Imperial Impact: Studies in the Economic History of Africa and India* (London, 1977).

Chattopadhyay, Chhanda. "The Growth of the 'Punjab Tradition' Emergence and Consolidation of Small Peasant Tenures in Nineteenth-Century Punjab", *Bengal Past and Present,* C IV (1985).

Chaudhuri, K.N. "Economic Problems and Indian Independence", in C.H. Philips and M.D. Wainwright (eds.), *Partition of India: Policies and Perspectives,* 1935-47 (London, 1970).

———. "Foreign Trade and Balance of Payments (1757-1947)", in Dharma Kumar (ed.), *The Cambridge Economic History of India,* Vol. 2, 1757-1970.

Chhabra, G.S. *The Advanced History of the Punjab* (Ludhiana, 1962).

———. *Social and Economic History of the Punjab,* 1849-1901 (Jullunder, 1962).

Chowdhury, Prem, *Punjab Politics: The Role of Chhotu Ram* (New Delhi, 1984).

———. "Agrarian Relations Prevailing in the Punjab at the Time of the Enactment of the So-called Golden Laws or Agrarian Legislations of the Late Thirties", *Punjab Past and Present,* V (1976).

Clairmonte, F.F. *Econoic Liberalism and Underdevelopment* (Bombay, 1960).

Cooper, Adrienne. *Sharecropping and Sharecroppers' Struggles in Bengal,* 1930-1950 (Calcutta).

Darling, M.L. *The Punjab Peasant in Prosperity and Debt,* 4th edn. (London, 1947).

———. *Wisdom and Waste in the Punjab Village* (London, 1934).

Davey, Brian. *The Economic Development of India* (Nottingham, 1975).

Dewey, C.J. "The Agricultural Statistics of the Punjab, 1867-1947", *Bulletin of Quantitative and Computer Methods in South Asian Studies,* 2 (1974).

———. "Patwari and Chaukidar: Sub-ordinate officials and Reliability of India's Agricultural Statistics", in C.J. Dewey and A.G. Hopkins (eds.), *The Imperial Impact: Studies in Economic History of Africa and India* (London, 1978).

———. (ed.), *Arrested Development in India: The Historical Dimension* (New Delhi, 1988).

Domin, Dolores. "Some Aspects of British Land Policy in Punjab after Its Annexation in 1849", *Punjab Past and Present,* VIII (1974).

Dungen, P.H.M. Van den. *The Punjab Tradition: Influence and Authority in Nineteenth Century India* (London, 1972).

Dutt, R.P. *India Today* (Bombay, 1949).

Dutt, Romesh. *The Economic History of India in the Victorian Age,* 6th edn. (London, 1956).

Emerson, H.A. *A Note on the Land System in the Punjab* (Lahore, 1938).

Emmanuel, A. *Unequal Exchange: A Study of the Imperialism of Trade* (New York, 1972).

Fagan, P.J. *Possibilities of Agricultural Development in the Punjab* (Lahore, 1922).

Farmer, B.H. *Agricultural Colonization in India since Independence* (London, 1974).

Fox, Richard G. *Lions of the Punjab* (Berkeley, 1985).

———. "Urban Class and Communal Consciousness in Colonial Punjab: The Genesis of India's Intermediate Regime", *Modern Asian Studies,* 18 (1984).

———. "British Colonialism and Punjabi Labour", in Charles Bergquist (ed.), *Labour in the Capitalist World Economy* (Beverly Hills, 1984).

Framji, K.K. and I.K. Mahajan. *Irrigation and Drainage in the World: A Global Review,* Vol. 1, 2nd edn. (New Delhi, 1969).

Frank, A.G. "The Development of Underdevelopment" in Robert Rhodes (ed.), *Imperialism and Underdevelopment: A Reader* (New York, 1970).

Galbraith, J.K and J.D. Black. "The Maintenance of Agricultural Production During Depression: Explanations Reviewed", *Journal of Political Economy*, XLVI.

Ghosh, A. "The Impact of Commercial Growth on Agricultural Tenure Systems in India", *The Manchester School*, 23, (1955).

Gibbling, M.T. *Concise Report on the Punjab Irrigation Works* (Bombay, 1932).

Giddens, Anthony. *Capitalism and Modern Social Theory* (Cambridge, 1971).

Gough, K. and H.P. Sharma. (eds.), *Imperialism and Revolution in South Asia* (New York, 1973).

Guha, Sumit. *The Agrarian Economy of the Bombay Deccan*, 1818-1941 (New Delhi, 1985).

Gustafson, W.E. and K.W. Jones. *Sources on Punjab History* (New Delhi, 1975).

Hamid, Naved. "Dispossession and Differentiation of the Peasantry in the Punjab During Colonial Rule", *Journal of Peasant Studies*, 10 (1982).

Harnetty, P. "Cotton Exports and Indian Agriculture, 1861-1870", *Economic History Review*, 2nd series, XXIV (1971).

Howard, Albert. *Crop Production in India: A Critical Survey of its Problems* (New Delhi, 1982).

Ibbetson, Denzil. *The Punjab Castes* (Lahore, 1916).

Islam, M. Mufakharul. "Trends in Crop Production in Undivided Punjab: A Reassessment" in C.J. Dewey (ed.), *Arrested Development in India: The Historical Dimension* (New Delhi, 1988).

———. *Bengal Agriculture, 1920-1946: A Quantitative Study* (Cambridge, 1974).

———. "M. L. Darling and the Punjab Peasant in Prosperity and Debt: A Fresh Look", *Journal of Peasant Studies*, 13 (1985).

———. "Irrigation and Crop Patterns in Punjab, 1887-1947: A Preliminary View", *Proceedings of International Conference on Indian Ocean Studies* (Perth, 1984).

———. "Irrigation and Punjab Agriculture, 1906-1945", *South Asia*, 1, 1 (1978).

———. "Economic Policy of the Indian Government in the Second Half of the Nineteenth Century: Laissez Faire or Selective Intervention?", in S.S. Akanda (ed.), *Studies in Modern Bengal*, I.B.S Seminar 2 (Rajshahi, 1981).

———. "Punjab Land Alienation Act and the Professional Moneylenders" *Modern Asian Studies* (forthcoming).

Kaul, Minoti Chakravarty. "The Commons, Community and the Courts of Colonial Punjab", *Indian Economic and Social History Review*, XXIV, No. 4 (December, 1992).

Kessinger, T.K. *Vilayatpur*, 1848 (Berkeley, 1974).

———. "The Peasant Farm in North India, 1848-1968", *Explorations in Economic History*, XII (1975).

Khosla, G.S. "The Growth of the Railway System in the Punjab", in Harbans Singh and N.G. Barrier (eds.), *Punjab Past and Present: Essays in Honour of Dr. Ganda Singh* (Patiala, 1976).

Kitching, G. *Development and Underdevelopment in Historical Perspective: Populism, Nationalism and Industrialization* (London, 1982).

Krishna, R. "Farm-Supply Response in India-Pakistan: A Case Study of the Punjab Region", *Economic Journal*, September 1963.

Krishnamurthy, T. "Changing Concepts of Work in the Indian Census: 1901-1961", *The Indian Economic and Social History Review*, XIV (1977).

Kumar, D. (ed.), *The Cambridge Economic History of India*, Vol.11 (Cambridge, 1984).

Kumar, N.K. "Hydraulic Agriculture in Peninsular India (*c.* 300 B.C. to 1300 A.D.)", *Proceedings of the Indian History Congress*, 1979 (New Delhi, 1980).

Kumar, Ravinder. *Essays in the Social History of Modern India* (Calcutta, 1983).

Kuznets, S. *et al.* (eds.), *Economic Growth: Brazil, India and Japan* (Durham NC., 1955).

Laclau, Ernesto. *Perspectives on Imperialism: Feudalism and Capitalism in Latin America* (New York, 1971).

Latif, A. *Industrial Punjab* (Bombay, 1911).

Laxminarayan, H. and S.S. Tyagi. "Inter-state Variations in Types of Tenancy", *Economic and Political Weekly*, Special Number, September 1977.

Lehman, David. "Two Paths of Agrarian Capitalism or a Critique of Chayanovian Marxism", *Comparative Studies in Society and History*, 28 (4).

Lipton, M. "The Theory of Optimising Peasant", *Journal of Development Studies*, Vol. 4, No. 3 (1968).

Ludden, David. *Peasant History in South India* (New Delhi, 1989).

Macpherson, W.J. "Economic Development in India under the British Crown, 1858-1947" in A.J. Youngson (ed.), *Economic Development in the Long Run* (London, 1972).

Madan, B.K. *Some Aspects of Rural Economy in the Punjab* (Lahore, 1934).

Maddison, A. *Class Structure and Economic Growth: India and Pakistan Since the Moghuls* (London, 1917).

Mann, Susan Archer. *Agrarian Capitalism in Theory and Practice* (Chapel Hill, 1990).

Marx, Karl and F. Engles. *On Colonialism* (Moscow, 1975).

Mathur, Y.B. *British Administration of Punjab*, 1849-1875 (New Delhi, 1966).

Meier, G.M. *The International Economics of Development* (Tokyo, 1968).

Meier, G.M. and R.E. Balwin. *Economic Development: Theory, History, Policy* (New York, 1966).

Mody, A. "Population Growth and Commercialization of Agriculture: India, 1890-1940", *Indian Economic and Social History Review*, 19.

Michel, A.A. *The Indus Rivers* (New Haven, 1967).

Minhas, B.A. and A. Vaidyanathan. "Growth of Crop Output in India, 1951-54 to 1958-61: An Analysis by Component Elements", *Journal of the India Society of Agricultural Statistics*, 17 (2).

Mishra, S.C. "Commercialization, Peasant Differentiation and Merchant Capital in Late Nineteenth-Century Bombay and Punjab", *Journal of Peasant Studies*, 10 (1982).

———. "On the Reliability of Pre-Independence Agricultural Statistics in Bombay and Punjab", *Indian Economic and Social History Review*, 20.

Mitra, Ashok. *Terms of Trade and Class Relations* (London, 1977).

Moore, B. *The Social Origins of Dictatorship and Democracy* (Boston, 1966)

Mouzelis, N. "Capitalism and the Development of Agriculture", *Journal of Peasant Studies*, 4 (7).

Mukerji, K. "Land Prices in Punja", in M.K. Chawdhuri (ed.), *Trends of Socio-Economic Change in India, 1871-1961* (Simla, 1969).

Mukherjee, Mridula. "Some Aspects of the Agrarian Structure of the Punjab, 1925-47", *Economic and Political Weekly*, Vol. XV, No. 26 (1980).

———. "Commercialization and Agrarian Change in Pre-Independence Punjab", in K.N. Raj *et al.* (eds.), *Essays on the Commercialization of Indian Agriculture* (Delhi, 1985).

Mukhia, Harbans. "Marx on Pre-colonial India: An Evaluation", in Diptendra Banerjee (ed.), *Marxian Theory and the Third World* (New Delhi, 1985).

Nair, Kusum. *In Defense of the Irrational Peasant* (Universtiy of Chicago Press, 1979).

Nakamura, J.I. *Agricultural Production and Economic Developement of Japan* (Princeton, 1966).

Narain, Brij. *Indian Economic Life: Past and Present* (Lahore, 1929).

Nijjar, B.S. *Punjab Under the British Rule* (1849-1947), Vol. III (New Delhi, 1974).

Ohkawa, K. "Concurrent Growth of Agriculture and Industry: A Study of Japanese Case", in R.N. Dixey (ed.), *International Explorations of Agricultural Economics* (Ames, Iowa, 1964).

Ohkawa, K. *et al.* (eds.), *Agriculture and Economic Growth: Japan's Experience* (Tokyo, 1969).

Oren, S. "The Sikhs, Congress and the Unionists in the British Punjab", *Modern Asian Studies,* VII (1974).

Ombedt, G. "Migration in Colonial India: The Articulation of Feudalism and Capitalism in Colonial India", *Journal of Peasant Studies,* 7 (1980).

Panse, V.G. *Report on the Scheme for the Improvement of Agricultural Statistics* (Simla, 1946).

Parker, W.N. (ed.), *Economic History and the Modern Economist* (London, 1986).

Patnaik, Usha. "Neo-Populism and Marxism: The Chaynovian View of the Agrarian Question and Its Fundamental Fallacy", *Journal of Peasant Studies,* 6 (1979).

Paustian, P.W. *Canal Irrigation in the Punjab* (New York, 1930).

Perlin, Frank. "Growth of Money Economy and Some Questions of Transition in Pre-Colonial India", *Journal of Peasant Studies,* 11 (1984).

Prakash, O.M. *Agrarian Legislation in the Punjab,* Vol. I (Lahore, 1940).

Pray, C.E. "Accuracy of Agricultural Statistics and Sources of Growth in the Punjab", *Indian Economic and Social History Review,* 21.

Puri, Nina. *Political Elite and Society in the Punjab* (New Delhi, 1985).

Raj, K.N. *et al.* (eds.), *Essays on the Commercialization of Indian Agriculture* (Delhi, 1985).

Ray, Rajat K. *Entrepreneurship and Industry in India,* 1800-1947 (New Delhi, 1992).

Roberts, Michael. "Irrigation Policy in British Ceylon During the Nineteenth Century", *South Asia,* 2 (1972).

Roberts, W. and S.B.S. Kartar Singh. *A Text-Book of Punjab Agriculture* (Lahore, 1947).

Rothermund, Dietmar. *An Economic History of India* (New York, 1988).

———. *The Indian Economy Under the British Rule* (New Delhi, 1983).

———. *India in the Great Depression, 1929-1939* (New Delhi, 1992).

Rudra, Ashok. *Indian Agricultural Economics: Myths and Realities* (New Delhi, 1982).

Rudra, Ashok. *et al. Studies in the Development of Capitalism in India* (Lahore, 1978).

Rungta, R.S. *Rise of Business Corporations in India,* 1851-1900 (Cambridge, 1970).

Saini, B.S. *The Social and Economic History of the Punjab* (Delhi, 1979).

Schultz, T. *Transforming Traditional Agriculture* (Yale, 1964).

Scott, J.C. *The Moral Economy of the Peasant* (Yale University Press, 1976).

Shah, K.T. *Sixty Years of Indian Finance,* 2nd edn. (Bombay, 1927).

———. *Industrialization of the Punjab* (Lahore, 1947).

———. *River Training and Irrigation* (Bombay, 1947).

Shanin, T. *Peasants and Peasant Societies* (Penguin, 1971).

———. "The Nature and Logic of the Peasant Economy, Part I", *Journal of Peasant Studies,* 1 (1).

———. "The Nature and Logic of the Peasant Economy, Part II, *Journal of Peasant Studies,* 1 (2).

Sharma, H.C. "Caste Mobility Among the Artisan Castes of the Punjab", *Proceedings of the Punjab History Conference,* 13th Session (March 1979).

Sharma, I. *Land Revenue Administration in the Punjab,* 1849-1901 (New Delhi, 1985).

Sharma, S.C. *Punjab: The Crucial Decade* (New Delhi, 1987).

Sarma, S.R. *Punjab in Ferment* (New Delhi, 1971).

Shigemochi, Hirashima. *The Structure of Disparity in a Developing Agriculture: A Case Study of the Pakistan Punjab* (Tokyo, 1974).

Singh, Chetan. "Well-irrigation Methods in Medieval Punjab: The Persian Wheel Reconsidered", *Indian Economic and Social History Review,* XXII (1985).

Singh, Fauja. "The Effects and Land Transfer on Rural Economy During the Latter Half of the 19th Century", *Proceedings of the Punjab History Conference*, March 1979.

Singh, Sukhwant. "Government Policy and Agricultural Development in Punjab, 1849-1901", *Proceedings of the Punjab History Conference*, March 1979.

Sinha, B. and R. Bhatia. *Economic Appraisal of Irrigation Projects in India* (New Delhi, 1982).

Stone, Ian. *Canal Irrigation In British India: Perspectives on Technological Change in a Peasant Society* (Cambridge, 1984).

Talbot, Ian. *Punjab and the Raj, 1849-1947* (New Delhi, 1988).

Tandan, P. *Punjabi Century, 1847-1947* (London, 1963).

Thavaraj, K. "Pattern of Public Investment in India, 1900-1939", *Indian Economic and Social History Review*, 1 (1963).

Thompson, W.P. *Punjab Irrigation* (Lahore, 1925).

Thornburn, S.S. *Mussalmans and Moneylenders in the Punjab*, rpt. (New Delhi, 1984).

Thornburn, S.S. *The Punjab in Peace and War* (Lahore, 1979).

Tomlinson, B.R. *The Political Economy of the Raj* (Cambridge, 1979).

———. "The Historical Roots of Indian Poverty: Issues in the Economic and Social History of Modern South Asia, 1880-1960", *Modern Asian Studies*, XXII (1988).

Tomosugi, Takashi. *Rethinking the Substantive Economy in Southeast Asia: On the Margins Between Utilities and Meanings* (Tokyo, 1991).

Trevaskis, H.K. *The Land of the Five Rivers* (London, 1934).

———. *The Punjab of Today*, 2 Vols (Lahore, 1931).

Ulyanovsky, Rostilav. *Agrarian India Between the World Wars* (Moscow, 1981).

Valkil, C.N. *Financial Developments in Modern India* (Bombay, 1925).

Vergopoulos, Kostas. "Capitalism and Peasant Productivity", *Journal of Peasant Studies*, 5 (4).

Whitcombe, E.M. *Agrarian Conditions in Northern India* (Berkely, 1972).

———. "Irrigation" in Dharma Kumar (ed.) *Cambridge Economic History of India*, Vol. 2.

Wiltfagel, C.A. *Oriental Despolism* (New Haven, 1957).

Thorner, Daniel. "Peasant Economy as a Category in Economic History", in Theodor Shamin (ed.), *Peasants and Peasant Societies*.

———. *The Agrarian Prospect in India* (New Delhi, 1956).

———. *Land and Labour in India* (Bombay, 1962).

Unakar, R.S.M.V. "Correlation Between Weather and Crops with Special Reference to Punjab Wheat", *Memoirs of the Metereological Department of India*, 25.

Unpublished Ph.D. Theses

Agnihotri, Indu. "Agrarian Change in the Canal Colonies, 1890-1935" (Jawaharlal Nehru University, 1987).

Ahmad, K.S. "The Agricultural Georaphy of the Punjab" (University of London, 1939).

Bhattacharya, Neeladri. "Agrarian Change in the Punjab, 1880-1940" (Jawaharlal Nehru University, 1985).

Gujral, S.S. "Population Growth and Issues in Indian Economic History, 1851-1901" (George Town, 1972).

Menezes, O.J. "Agricultural Stagnation as an Obstacle to Industrial Growth in India, 1920-50" (Princeton, 1958).

Misra, S.C. "Patterns of Long Run Agrarian Change in Bombay and Punjab", 1881-1972 (Cambridge, 1981).

Pray, C.E. "Economics of Agricultural Research in British Punjab and Pakistan Punjab" (Pennsylvania, 1978).

Saini, K.G. "Some Measures of Economic Growth of India: 1860-1913" (Columbia, 1968).

Sen, Chiranjib. "Essays on the Transformation of India's Agrarian Economy" (Stanford, 1978).

Siddiqui, A.J. "Canal Irrigation, Agricultural Development and Social Change in West Pakistan, 1880-1965" (Wisconsin, 1965).

Zafar, Fareeha. "The Impact of Canal Construction on the Rural Structures of the Punjab: The Canal Colony Districts, 1880-1940".

INDEX